Managing Reality

Book Five

Managing Procedures

Bronwyn Mitchell and Barry Trebes

Published by Thomas Telford Publishing, Thomas Telford Ltd, 1 Heron Quay, London E14 4JD. URL: www.thomastelford.com

Distributors for Thomas Telford books are
USA: ASCE Press, 1801 Alexander Bell Drive, Reston, VA 20191-4400, USA
Japan: Maruzen Co. Ltd, Book Department, 3–10 Nihonbashi 2-chome, Chuo-ku, Tokyo 103
Australia: DA Books and Journals, 648 Whitehorse Road, Mitcham 3132, Victoria

First published 2005

Also available in this series from Thomas Telford Books
NEC – Managing Reality: Introduction to the Engineering and Construction Contract. ISBN 07277 3392 3
NEC – Managing Reality: Procuring an Engineering and Construction Contract. ISBN 07277 3393 1
NEC – Managing Reality: Managing the contract. ISBN 07277 3394 X
NEC – Managing Reality: Managing change. ISBN 07277 3395 8
NEC – Managing Reality: Complete box set. ISBN 07277 3397 4

Also available from Thomas Telford Books
NEC3 (complete box set). ISBN 07277 3382 6

A catalogue record for this book is available from the British Library

9 8 7 6 5 4 3 2 1

ISBN: 0 7277 3396 6

In this document, except where the context shows otherwise, words in the singular also mean in the plural and the other way around and words in the masculine also mean in the feminine and neuter.

Typeset by Academic + Technical, Bristol
Printed and bound in Great Britain by Bell & Bain Limited, Glasgow, UK

Preface

Now more than a decade on from its initial formal introduction (1st edition 1993), the NEC form of contract remains radical in its ethos and contemporary in its management principles for delivering successful projects in the business environment of 21st century construction. Although conceived in the mid-1980s, in what could be described as a decade of success and excess, with a construction industry racked by conflict and confrontation, the NEC owes much of its current widespread and growing usage to the deep recession of the early 1990s, which forced the construction industry to rethink its approach and performance, as much from a need to survive as from a desire to improve.

It is generally recognised and accepted that Sir Michael Latham's *Constructing the Team*, published in 1994, and Sir John Egan's *Rethinking Construction*, published in 1998, were the two main catalysts and energisers of change and improvement within the construction industry throughout the last decade of the 20th century and into the 21st century. Arguably the NEC was the third key driver of cultural reform and management discipline; indeed the NEC was formally recognised by Latham as being the contract form which, more than any other, aligned to his vision for future construction and Egan similarly embraced the NEC as part of his movement for innovation.

Usage has brought with it practical experience and accordingly the thrust of this book is about dealing with the reality of real-life projects. The authors have a combined knowledge and hands-on experience of NEC-managed projects spanning some 20 years and have therefore written the book with the specific purpose of advising and assisting those who would wish to use or even those who already do use the ECC, on the concepts of modern contract practices, procedures and administration.

This book is about 'how to': how to manage the ECC contract and how to administer it. As such, it does not attempt to give a legal treatise or a blow-by-blow review of each and every clause and certainly is not a rehash of the NEC/ECC Guidance Notes. It is intended to be complementary to other publications, which give excellent theoretical and legal perspectives. This book is about managing reality.

Although now regarded as a 'mature' form of contract the ECC is still relatively young when compared to more traditional forms. Therefore, even with current practitioners, experience will vary both in duration and depth. With this in mind the book has been consciously structured so as to be presented as a five-part book-set that covers the needs of the student professional or prospective client, through to the novice practitioner and experienced user. It provides a rounded view of the ECC, whatever your discipline, on both sides of the contractual relationship and is aimed at enabling everyone to realise the business benefits from using the NEC suite of contracts generally and the ECC in particular.

Since the NEC's official launch in 1993, adoption and usage of the NEC, renamed *Engineering and Construction Contract – ECC* in 1995 (2nd edition), has grown, such that it is now the most frequently used contract form for civils, transportation infrastructure and utilities works and is increasingly the preferred form for building construction projects. The latest edition of the contract (NEC3) was issued in July 2005.

The wide acceptance of the NEC generally and its elevation in stature as the preferred contract form is further reinforced by the Office of Government Commerce's endorsement of the third edition (NEC3). These five books take account of the changes introduced to the contract within NEC3.

Foreword

As almost the first UK client to use the NEC, even in its consultative form prior to its launch in 1993, I believe *NEC: Managing Reality* to be a welcome addition to the construction bookshelf: an essential introduction to the NEC for the prospective 'novice' practitioner and an excellent *aide-mémoire* reference book for the regular user of this form of contract.

The joy of this new book is that it brings together a wealth of practical expertise and knowledge from two of the UK's most experienced NEC practitioners written in a style that keeps faith with NEC principles of clarity and simplicity, while respecting differing levels of knowledge within its potential readership.

I have always believed that choosing the contract form is as much a business issue as a construction decision and that business success only results from good management. Certainly this book gives emphasis to the ethos of managing for success rather than the reactive debate of failure. It imparts knowledge, understanding and practical experience of the Contract in use and equally stresses the roles, responsibilities and discipline of the management procedures that apply to everyone. I particularly like the five-module format, in that it presents itself in readable, manageable chunks that can be readily digested or revisited whatever the reader's previous experience of its use.

This book educates, giving guidance and confidence to anyone dealing with real contract issues, following both the spirit as well as the letter of the Contract. It is arguably the most comprehensive practical treatise to date on how to manage and administer the ECC. It is a must for the professional office. Every 'home' should have one.

David H Williams, CEng, FICE
Chairman, Needlemans

(*Formerly Group Construction and Engineering Director BAA plc; founding Chairman, NEC UK Users Group: 1994–1997*)

Introduction

General

This series of books will provide the people who are actually using the Engineering and Construction Contract (ECC) in particular, and the New Engineering Contract (NEC) suite in general, practical guidance as to how to prepare and manage an ECC contract with confidence and knowledge of the effects of their actions on the Contract and the other parties.

Each book in the series addresses a different area of the management of an ECC contract.

- Book One – NEC Managing Reality: Introduction to the Engineering and Construction Contract
- Book Two – NEC Managing Reality: Procuring an Engineering and Construction Contract
- Book Three – NEC Managing Reality: Managing the Contract
- Book Four – NEC Managing Reality: Managing Change
- Book Five – NEC Managing Reality: Managing Procedures

- *Book One (NEC Managing Reality: Introduction to the Engineering and Construction Contract)* is for those who are considering using the ECC but need further information, or those who are already using the ECC but need further insight into its rationale. It therefore focuses on the fundamental cultural changes and mind-shift that is required to successfully manage the practicalities of the ECC in use.

- *Book Two (NEC Managing Reality: Procuring an Engineering and Construction Contract)* is for those who need to know how to procure an ECC contract. It covers in practical detail the invitations to tender, evaluation of submissions, which option to select, how to complete the Contract Data and how to prepare the Works Information. The use of this guidance is appropriate for employers, contractors (including subcontractors) and construction professionals generally.

- *Book Three (NEC Managing Reality: Managing the Contract)* is essentially for those who use the contract on a daily basis, covering the detail of practical management such as paying the contractor, reviewing the programme, ensuring the quality of the works and dispute resolution. Both first-time and experienced practitioners will benefit from this book.

- *Book Four (NEC Managing Reality: Managing Change)* is for those who are managing change under the contract; whether for the employer or the contractor (or subcontractor) the management of change is often a major challenge whatever the form of contract. The ECC deals with change in a different way to other more traditional forms. This book sets out the steps to efficiently and effectively manage change, bridging the gap between theory and practice.

- *Book Five (NEC Managing Reality: Managing Procedures)* gives step-by-step guidance on how to apply the most commonly used procedures, detailing the actions needed by all parties to comply with the contract. Anyone administering the contract will benefit from this book.

Background

The ECC is the first of what could be termed a 'modern contract' in that it seeks to holistically align the setting up of a contract to match business needs as opposed to writing a contract that merely administers construction events.

The whole ethos of the ECC, or indeed the NEC suite generally, is one of simplicity of language and clarity of requirement. It is important that the roles and responsibilities are equally clear in definition and ownership.

When looking at the ECC for the first time it is very easy to believe that it is relatively straightforward and simple. However, this apparent simplicity belies the need for the people involved to think about their project and their role and how the ECC can deliver their particular contract strategy.

The ECC provides a structured flexible framework for setting up an appropriate form of contract whatever the selected procurement route. The fundamental requirements are as follows.

- The Works Information – quality and completeness – what are you asking the Contractor to do?
- The Site Information – what are the site conditions the Contractor will find?
- The Contract Data – key objectives for completion, for example start date, completion date, programme – when do you want it completed?

The details contained in the series of books will underline the relevance and importance of the above three fundamental requirements.

The structure of the books

Each chapter starts with a synopsis of what is included in that chapter. Throughout the book there are shaded 'practical tip' boxes that immediately point the user towards important reminders for using the ECC (see example below).

> Clarity and completeness of the Works Information is fundamental.

There are also unshaded boxes that include examples to illustrate the text (see example below).

> Imagine a situation in which the *Supervisor* notifies the *Contractor* that the reinstatement of carriageways on a utility diversion project is not to the highway authority's usual standards. However, the Works Information is silent about the reinstatement.
>
> Although it is not to the authority's usual standard, it is **not** a Defect because the test of a Defect is non-conformance with the Works Information. In this situation, if the *works* need to be redone to meet the authority's requirements, the *Contractor* is entitled to a compensation event because the new requirements are a change to the Works Information.

Other diagrams and tables are designed to maintain interest and provide another medium of explanation. There are also standard forms for use in the administration and management of the contract together with examples.

Throughout the books, the following terms have been used in a specific way.

- NEC is the abbreviation for the suite of New Engineering Contracts and it is not the name of any single contract.
- ECC is the abbreviation for the contract in the NEC suite called the Engineering and Construction Contract.

The NEC suite currently comprises the

- Engineering and Construction Contract
- Engineering and Construction Subcontract

- Engineering and Construction Short Contract
- Engineering and Construction Short Subcontract
- Professional Services Contract
- Adjudicator's Contract
- Term Service Contract
- Framework Contract

Acknowledgements

We would like to thank the following individuals and companies who have supported the book.

Andy Door who gave advice on procurement within the public sector. Mike Attridge of Ellenbrook Consulting who reviewed the book on behalf of the authors. David H. Williams who provided guidance and support in the development of the book and everyone at Needlemans Construction Consultants and MPS Limited.

Series Contents

The following outlines the content of the five books in the series.

Book 1 NEC Managing Reality: Introduction to the Engineering and Construction Contract

Preface, Foreword, Introduction and Acknowledgements
Series Contents, Contents, List of tables, List of figures

Chapter 1 Introduction to the Engineering and Construction Contract, concepts and terminology

Synopsis
This chapter looks at:

- An introduction to the ECC
- An identification of some of the differences between the ECC and other contracts
- A brief outline of differences between ECC2 and ECC3
- An outline of the key features of the ECC
- Conventions of the ECC
- Concepts on which the ECC is based
- Terminology used in the ECC
- Terminology not used in the ECC
- How the ECC affects the way you work

Appendix 1 Summary of differences between ECC2 and ECC3

Chapter 2 Roles in the Engineering and Construction Contract

Synopsis
This chapter describes the roles adopted in the ECC including:

- How to designate a role
- Discussion of the roles described in the ECC
- Discussion of the project team
- How the ECC affects each of the roles

Appendix 2 List of duties

Book 2 NEC Managing Reality: Procuring an Engineering and Construction Contract

Preface, Foreword, Introduction and Acknowledgements
Series Contents, Contents, List of tables, List of figures

Chapter 1 Procurement

Synopsis
This chapter looks at the concept of procurement and contracting strategies and discusses:

- Procurement and contract strategy
- What tender documents to include in an ECC invitation to tender
- How to draft and compile a contract using the ECC
- Procurement scenarios that an employer could face and how to approach them

Book 5 NEC Managing Reality: Managing Procedures

Preface, Foreword, Introduction and Acknowledgements
Series Contents, Contents, List of tables, List of figures

Chapter 1 ECC Management: Procedures

Synopsis
This chapter brings together all the aspects discussed in previous chapters in Books 1 to 4, which form part of the series of books on NEC Managing Reality. This chapter provides the 'how to' part of the series. It introduces some example pro-formas for use on the contract.

For quick reference, this chapter may be read on its own. It does not, however, detail the reasons for carrying out the actions, or the clause numbers that should be referred to in order to verify the actions in accordance with the contract. These are described in detail in other chapters that form part of this series.

Contents

List of figures

1 ECC Management: Procedures

Synopsis

This chapter brings together all the aspects discussed in previous chapters in Books 1 to 4, which form part of the series of books on NEC Managing Reality. This chapter provides the 'how to' part of the series. It introduces some example pro-formas for use on the contract. Unless detailed separately due to a complex procedure, replies are described under the relevant action.

For quick reference, this chapter may be read on its own. It does not, however, detail the reasons for carrying out the actions, or the clause numbers that should be referred to in order to verify the actions in accordance with the contract. These are described in detail in other chapters that form part of this series.

1.1 Introduction

Each employer and contractor will generally have his own methods and systems for contract administration. If the ECC is being used for the first time, it is vital to ensure that the procedures are understood and that the systems are altered to take into account any new requirements of the ECC. One small aspect of this is the pro-formas that may be used by both parties in administering the contract.

The *Employer* should ensure that the *Supervisor* and the *Project Manager* are quite aware of their duties. Both the *Supervisor* and the *Project Manager* should sit down and discuss how they intend to approach the contract and how they intend to use the systems put in place[1] so that they are in consensus as to the general approach to the contract.

1.2 Records to maintain

The *Employer* and his *Project Manager* should also ensure that documentary evidence, such as site diaries, is considered. The *Contractor* should ensure that plant and labour records, allocation sheets, time sheets and other documentary evidence required for payment purposes are established and maintained.

Both Parties should establish and maintain a commercial filing system on site in which to keep the required records.

1.2.1 The importance of good site records

Good site records are an essential tool in the management of time and it is essential that the *Supervisor* realises the importance of keeping good site diaries and how this information assists in the overall management of the contract.

The people involved in the contract are encouraged to keep jointly agreed site diaries which record what happened each day.

It is essential that a daily site diary is maintained for each geographical area of a contract, for example work area A, B or C, etc.

The site diary can be invaluable in establishing the events that took place on any particular day, particularly in the case of compensation events about causes of and responsibility for delay and/or disruption.

On a more general level they serve to keep the project team aware of day-to-day issues, alerting them to problems (potential or current) with commercial implications. The diaries should be considered as being a supplement to regular site visits by the project team responsible for the various geographical locations.

To get the maximum benefit from the diary, it is important that only 'pertinent information' is recorded, and that this is noted in a concise and dispassionate manner. As guidance in preparing the site diary, listed below is the type of information that it may be necessary to record. The list is by no means exhaustive, the emphasis being on the judgement of the diary-keeper to decide what information is pertinent:

- weather conditions,
- site conditions,
- the principal locations where work was being undertaken together with principal resources being utilised and if possible an indication of output achieved,
- any delays noted (for whatever reason), for example weather, plant breakdown, late instructions, damage to third-party property, existing services discovered, inefficiencies,
- any verbal instructions given,
- any complaints by third parties,
- any work by others at the site, for example other package contractors,

[1] For example, if they intend to copy all correspondence to each other, or just the certificates that they are obliged to under the contract (clause 13.6).

- any verbal notification by the *Contractor*'s site staff regarding difficulties encountered.

A 'typical' diary is included below for your reference. (This is typed for clarity but hand-written diaries will obviously serve the purpose.)

Contract title:	Tunnel and Station Works		
Contractor:	ABC Ltd	**Contract Ref:**	06/05
Date:	2 May 2006	**Geographical area:**	Station
Weather:	Persistent rain all day	**Site conditions:**	Very wet

Work locations:

Station tunnel – excavation of top heading (approx. 2 metres advanced) – Liebherr 932, Cat 915 Shovel, Shotcrete machine and 2 miners.

Vent Shaft – continuing to sink 'soft piles' (female) to shaft (6 No. completed) – Cassegrade C30 (name of contractor).

Escape Shaft – continuing to sink shaft (2 No. rings today – 8 No. now in total) mobile crane (telescopic), mini-excavator.

Delays:

Obstruction encountered with piling to Vent Shaft – delay approx. 2 hours – suspected buried concrete.

Excavation for Escape Shaft progressing slowly most of the day due to mechanical problems with excavator – fitter arrived 3 pm.

Verbal instructions/notices:

ABC Ltd complained that Contractor G continued presence within the Vent Shaft worksite was disrupting their piling operations.

ABC Ltd advised that tests on samples taken of groundwater from the Escape Shaft showed contamination levels in excess of those anticipated.

Level B3 instructed not to commence male (hard) piles to Vent Shaft pending issue of revised loading criteria (expected from designer tomorrow).

Miscellaneous:

Complaints received at 4 pm re excessive noise/vibrations being experienced in Subway box – complaints coincided with time that obstruction was encountered. Employer's operational staff complained of muck wagons queuing outside Escape Shaft worksite – problem due to mechanical problems with excavator in shaft – will resolve itself now machine repaired.

Signed: ....

1.3 Review meetings

Normally meetings are held on a regular basis, perhaps weekly or monthly, to discuss progress. Other meetings, or as a separate agenda item within progress meetings, could be held to discuss early warnings and compensation events, how the programme has been affected, and possibly the budget in an Option C, D or E contract.

1.4 Contract administration

It is vital in any contract to ensure that correspondence is replied to swiftly and with as much information as possible. This becomes even more important in the ECC where the *period for reply* dictates the default period of time within which the *Project Manager* and the *Contractor* are obliged to reply to a communication one from the other. Some parts of the contract, such as the compensation event procedure, contain other time periods that are to be adhered to.

The *Contractor* is entitled to an assessment of time and money (as a compensation event) if the *Supervisor* or the *Project Manager* does not reply within the time period allowed in the contract. The *Supervisor* and the *Project Manager* should therefore ensure that they reply within the time required by the contract.

Although there is some sanction on the *Contractor* for tardiness,[2] the contract assumes in general that the *Contractor* will reply in time since it is clearly in his interests to do so. The three primary sanctions are the most important, however, since they affect the *Contractor*'s own budget and programme.

(1) Early warnings.
If the *Contractor* does not give an early warning that an experienced contractor could have given, then a compensation event is assessed as if he had given an early warning.[3] In other words, the *Contractor* is forced back into the position he would have been in had he carried out his obligations under the contract. His obligation is to notify an early warning **immediately** he becomes aware of any matter that could affect the budget, the programme or the performance of the project.[4]

(2) The programme.
If the *Contractor* does not submit a first programme for acceptance, then 25% of the amount due to him will be withheld until he does submit a first programme.[5] The *Contractor*'s obligation is to submit a first programme for acceptance at tender stage **or** within a certain period after the Contract Date, depending on the *Employer*'s requirement as stated in Contract Data part two or Contract Data part one respectively.

In addition, if there is no Accepted Programme, the *Project Manager* may assess a compensation event using his own assessment of the programme.[6]

If the *Contractor* has not submitted a revised programme (or, in ECC3, alterations to a programme) for acceptance as required by the contract, the *Project Manager* may assess a compensation event using his own assessment of the programme.[7]

(3) Compensation events.
In ECC2, if the *Contractor* does not notify a compensation event within **two weeks** of becoming aware of it,[8] he loses his contractual right to compensation and will therefore have to bear the burden of any delay or increase in cost as a result (with a recourse to adjudication). In ECC3, the *Contractor* has **eight weeks** of becoming aware of the compensation event to notify it to the *Project Manager*. If he does not do it within this time frame, then he is not entitled to changes to the Prices or programme unless the *Project Manager* should have notified the event to the *Contractor* but did not.

Therefore, in order to project manage the project effectively, the *Contractor* is advised to read and understand the contract and to submit communications in time.

1.4.1 Administration letters Examples of the contract administration letters that the *Project Manager*, *Contractor*, *Supervisor* and *Employer* could use throughout the period of the contract are contained within each section of Chapter 1.

1.5 Practical administration

1.5.1 Resourcing The ECC is a 'real-time' contract: it requires action today not tomorrow. The ECC is challenging in terms of the administration of communications and resources.

[2]Clause 16.1, 31.1, 61.3.
[3]ECC2 clauses 61.5 and 63.4; ECC3 clauses 61.5 and 63.5.
[4]Clause 16.1.
[5]Clause 50.3.
[6]Clause 64.2.
[7]Clause 64.2.
[8]Clause 61.3.

The level of the types of communication under the contract will be very dependent on the quality of the Works Information.

In pricing a tender the *Contractor*/Subcontractor will be required to decide on the resource levels required. This is very difficult to determine for the contract (the tender resource levels form part of the tendered target cost).

The *Contractor* takes the risk on his resource levels being correct. The *Contractor*/Subcontractor therefore needs to consider the likely levels of administration, compensation events, etc. This may need to be based on knowledge and experience of:

- the *Employer*,
- the *Consultant*,
- the particular project, for example refurbishment.

Compensation events can be like a No. 52 bus: there will be periods where there aren't any, and then several arrive together. This means that there are 'peaks and troughs' of activity which either require fluctuating resource levels or may be smoothed by extending the *period for reply* (see Fig. 1.1).

There is no opportunity to review this resource level if it is incorrect.

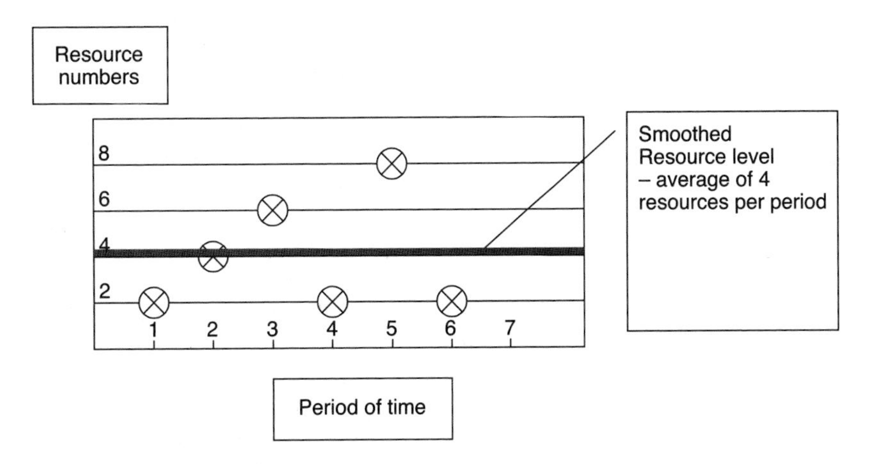

Fig. 1.1. Resource smoothing

1.5.2 Preparing the tender – *Contractor*/Subcontractor

When preparing the tender as a *Contractor*/Subcontractor there are a number of key points to remember.

(1) Realise that the ECC is different to other contracts.
(2) Ensure that you understand the Schedule of Cost Components.
(3) Ensure that you complete Contract Data part two. Do not be tempted to leave gaps.
(4) On the priced-based Options A and B ensure that you have allowed adequate resources for the 'real-time' management of the contract.
(5) Ensure that you have in your team the necessary programming skills and commercial/estimating skills.
(6) On priced-based contracts make sure the tendered *fee percentage* (ECC2) or *subcontracted fee percentage* and *direct fee percentage* (ECC3) are compatible with your own.
(7) Find out what the *Employer*'s policy is on compensation events which introduce work not in the original contract, for example specialist subcontract work. Will the *Contractor* be able to put forward the Subcontractor's quote? This is particularly relevant for Options A and B.
(8) What is the *Employer*'s policy on proposed instructions? Will this be used sparingly? How does the *Contractor*/Subcontractor know that this will not be abused? What about in situations where design work is called for?

(9) Ensure that your Subcontractor/sub-subcontractor understands the ECC. Ensure that the time-scales for the *period for reply* work within your own *period for reply.*

(10) Ensure that the project procedures are clear.

See also the items included within NEC Book 1, Managing Reality: Introduction to the Engineering and Construction Contract, Chapter 2.

1.5.3 Using technology to support ECC administration

The structure and procedures of the ECC lend themselves to the use of technology to support its administration. Its proactive style demands communication, early warning and resolution of problems to deliver an up-to-date business picture for all parties.[9]

The most effective and transparent way to achieve this, which minimises the 'paper chase' inherent in more traditional contracts, is to use technology-based process systems such as Management Process System Limited's ECC Contract Change Management (CCM) system.

The benefits that these technology-based systems can bring in relation to contract administration are:

- greater efficiency,
- clear audit trail,
- management which transcends location,
- data capture,
- performance data.

1.6 How to manage an ECC

This next section lists some of the actions that the various roles under the contract might be expected to carry out. For each action, a list of considerations is included. For some actions, example pro-formas are included for use with the ECC.

1.6.1 How to delegate an action

Both the *Project Manager* and the *Supervisor* may delegate any of their actions.[10]

This may be a relief to some employers whose project managers are project managers on many projects. However, it is preferable in ECC contracts that the *Project Manager* is dedicated to the project, particularly for larger projects, since the *Project Manager* should be on Site and aware of all that is affecting the project. Naming the same *Project Manager* on many projects is therefore not recommended.

It is possible, however, and may sometimes be necessary, for project managers to delegate their duties to other people, but preferably not the *Supervisor*. The *Supervisor* should always remain independent of the *Project Manager* and therefore should not be required to adopt some of his duties. It is confusing for the *Contractor* and may be confusing for the *Supervisor* since he has to remember which hat he is wearing in giving instructions or taking decisions. In particular, certain actions, such as acceleration, changing the Works Information and assessing the amount due, should not be delegated to the *Supervisor*.

Similarly, the *Supervisor* should not delegate any of his actions to the *Project Manager*.

Before any actions are delegated, the delegating party must first inform the *Contractor* of such delegation. It is also wise to notify for how long the delegation will be in place and also exactly what actions are being delegated. Note that a delegation does not prevent the *Project Manager* or the *Supervisor* from carrying out that duty himself. Close communication is therefore required. The *Project Manager* and the *Supervisor* can also cancel any delegation.[11]

[9] 'Choosing the contract form: a project or business decision?', article by David H. Williams Non-Executive Chairman of Needlemans in issue 17 of the NEC Users' Group newsletter dated April 2001.
[10] Clause 14.2.
[11] Clause 14.2.

The person to whom actions are delegated does not need to acquire a specific title under the contract, such as *Project Manager*'s Representative, since he will be the *Project Manager* for the actions that he is carrying out, and similarly for the *Supervisor*.

1.6.1.1 How the Project Manager *delegates an action*

(1) The *Project Manager* realises that he cannot perform all his duties under the contract because he does not have sufficient time to pay attention to all things happening under the contract and therefore he cannot be effective.
(2) The *Project Manager* decides to delegate some of his actions.
(3) He knows he should not delegate any actions to the *Supervisor*.
(4) He knows that he can still perform those actions that he has delegated.
(5) He knows he should not delegate some special actions such as acceleration, changing the Works Information and determining how much the *Contractor* should be paid.
(6) He decides which actions he can delegate and still remain in control of the management of the project.
(7) He decides to whom to delegate the actions and for how long.
(8) The *Project Manager* notifies the *Contractor* that he will be delegating some of his actions. He describes to whom he will be delegating the actions and for what period of time.
(9) The delegation is effective.

1.6.1.2 How the Project Manager *cancels a delegation*

The *Project Manager* may cancel a delegation at any time. Because the initial delegation notification would have had a time limit placed on it, the cancellation of a delegation need not be the issue of a cancellation notification; the delegation lapses automatically by virtue of time passing.

Where, however, the delegation is being cancelled while the initial delegation is still active, the *Project Manager* should issue a cancellation of delegation notification which mirrors the initial delegation, but which cancels the delegation. Where the delegation being cancelled was not time bound and therefore did not have an end date, the *Project Manager* is required to issue a delegation cancellation notification if he wishes to cancel the delegation.

1.6.1.3 How the Supervisor *delegates an action*

(1) The *Supervisor* is going on holiday and wishes to advise the *Contractor* that someone else will be carrying out his duties under the contract while he is off-site.
(2) He knows he should not delegate any actions to the *Project Manager*.
(3) He knows that he can still perform those actions that he has delegated, if he should decide to visit the Site during his holiday.
(4) He decides to whom to delegate the actions and calculates the duration.
(5) The *Supervisor* notifies the *Contractor* that he will be delegating all of his actions. He describes to whom he will be delegating the actions and for what period of time.
(6) The delegation is effective.

1.6.1.4 How the Supervisor *cancels a delegation*

The *Supervisor* may cancel a delegation at any time. He does so in the same manner as the *Project Manager*.

Pro-forma 1: DELEGATION OF THE *PROJECT MANAGER*'S DUTIES

Project number:

Description:

Contract number:

To: (The *Contractor*)	
Address:	
Telephone:	
Fax:	
Attention:	

In accordance with the terms of clause 14.2 of the Conditions of Contract I delegate the following *Project Manager*'s duties to:

Name:	
For the period:	

Clause No.	Duty

***Delegate*:**

Signature Name Date

***Project Manager*:**

Signature Name Date

Distribution:				

Pro-forma 2: DELEGATION OF THE *SUPERVISOR*'S DUTIES

Project number:

Description:

Contract number:

To: (The *Contractor*)	
Address:	
Telephone:	
Fax:	
Attention:	

In accordance with the terms of clause 14.2 of the Conditions of Contract I delegate the following *Supervisor*'s duties to:

Name:	
For the period:	

Clause No.	Duty

***Delegate*:**

Signature Name Date

***Supervisor*:**

Signature Name Date

Distribution:				

Pro-forma 3: DELEGATION OF THE *PROJECT MANAGER*'S DUTIES – CANCELLATION

Project number:

Description:

Contract number:

To: (The *Contractor*)	
Address:	
Telephone:	
Fax:	
Attention:	

In accordance with the terms of clause 14.2 of the Conditions of Contract I hereby cancel the previous delegation of the following *Project Manager*'s duties to:

Name:	

Clause No.	Duty

***Delegate*:**

Signature Name Date

***Project Manager*:**

Signature Name Date

Distribution:				

Pro-forma 4: DELEGATION OF THE *SUPERVISOR*'S DUTIES – CANCELLATION

Project number:

Description:

Contract number:

To: (The *Contractor*)	
Address:	
Telephone:	
Fax:	
Attention:	

In accordance with the terms of clause 14.2 of the Conditions of Contract I hereby cancel the previous delegation of the following *Supervisor*'s duties to:

Name:	

Clause No.	Duty

Delegate:

Signature Name Date

Supervisor:

Signature Name Date

Distribution:				

1.6.2 How to change the Works Information

1.6.2.1 How the Project Manager *changes the Works Information*

The *Project Manager* is the only person who may change the Works Information.[12] In particular, the *Supervisor* may not change the Works Information. Since all instructions are required to be in writing, the *Project Manager* effects a change to the Works Information by giving the *Contractor* a written instruction of this change.

Scenario 1
The *Supervisor* notices a Defect in one of the structures being erected by the *Contractor* and includes in his Defect notification a suggestion of how to change the structure so that the Defect is removed.

In this case, the *Contractor* would take note of the Defect and the requirement to correct it. He does not have to take on board any suggestions by the *Supervisor* that could be deemed to be changes to the Works Information since the *Supervisor* has no authority to change the Works Information. The *Contractor*'s obligation is to Provide the Works in accordance with the Works Information.

Scenario 2
The *Contractor*, while erecting the structure, realises that by a simple change he could improve the structure in a big way. He makes the change and continues to build the structure.

In this scenario, the *Contractor* has made a change to the Works Information. He has not fulfilled his contractual obligations and the change could, in fact, be regarded as a Defect. The *Contractor* should have taken his suggestion to the *Project Manager* and waited for an instruction to change the Works Information if the *Project Manager* considered that the change still met the *Employer*'s objectives.

Scenario 3
Two weeks after the *Contractor* has started on site, the *Project Manager* issues a drawing superseding a previous drawing.

This drawing issue is a change to the Works Information and the *Contractor* would be advised to notify a compensation event if the *Project Manager* does not do so. It may be that the change to the drawing is so small that no time or financial consequences arise, but this does not alter the fact that a change to the Works Information is a compensation event.

Scenario 4
The managing director of the client for whom a new office block is being built walks around the site and orders the *Contractor* to change the location of a supporting column because it is blocking his view of the lake from his new office. Since the *Project Manager* is the only person who has the authority to change the Works Information, the *Contractor* would be advised to inform the *Project Manager* of the managing director's wishes but not to make the change unless and until the *Project Manager* so orders.

[12] Clause 14.3.

Pro-forma 5: *PROJECT MANAGER*'S INSTRUCTION

Project number:

Description:

Contract number:

To: (The *Contractor*)	
Address:	
Telephone:	
Fax:	
Attention:	

In accordance with the terms of clause 14.3 the Works Information is changed as follows:

Delete:	
Add:	
Amend:	

		Tick
This change is a compensation event under clause 60.1(1).		
This is/is not	A change made in order to accept a Defect.	
This is/is not	A change to the Works Information requested by the *Contractor* for his design made:	
	• at his request or	
	• to comply with other Works Information provided by the *Employer*.	
I hereby instruct you to submit quotations for this compensation event.		
I hereby instruct you to submit alternative quotations for this compensation event.		
I hereby notify you that you did not give an early warning of this event which an experienced contractor could have given.		
I hereby state assumptions about the event because the effects of the compensation event are too uncertain to be forecast reasonably.		
You are not required to submit quotations for this event because:		
• the event arises from a fault of the *Contractor*,		
• quotations have already been submitted.		

***Project Manager*:**

Signature Name Date

Distribution:				

1.6.3 How to make changes to the people working on the contract

1.6.3.1 How the Employer *changes the* Project Manager

The *Employer* may replace the *Project Manager* or the *Supervisor* after he has notified the *Contractor* of the name of the replacement.[13]

(1) The *Project Manager* has been offered a better position with another company and has resigned from his current position. The *Employer* is required to replace the *Project Manager*.
(2) The *Employer* decides who will replace the *Project Manager*. If the *Project Manager* is in fact a member of staff in a consulting engineers firm, then this decision would be made by the firm, but it is the *Employer* from whom the notification should be addressed.
(3) The *Employer* notifies the *Contractor*, in writing, of the name of the person who will replace the *Project Manager* and the effective date of his appointment.
(4) The *Employer* replaces the *Project Manager*.

1.6.3.2 How the Employer *changes the* Supervisor

The *Employer* may replace the *Project Manager* or the *Supervisor* after he has notified the *Contractor* of the name of the replacement.[14] The procedure is the same as that for the *Project Manager*, above.

1.6.3.3 How the Contractor *notifies a replacement employee*

The *Contractor* is required to state the name, job, responsibilities, qualifications, and experience of each key person whom he employs on the project in Contract Data part two. This is to ensure that either those people work on the project or people with similar or better qualifications and experience replace them. If the *Contractor* wishes to replace a key person, he has to first submit the details of the proposed replacement to the *Project Manager* for acceptance and the *Project Manager* determines whether the qualifications and experience of the proposed replacement are as good as those of the key person before accepting or not accepting the proposed replacement.[15]

If the *Project Manager* has worked with the proposed replacement in the past and has experienced a personality clash, then he should either take his chances on the relevant compensation event clause 60.1(9) and withhold acceptance for this reason or state that the experience of the proposed replacement is not conducive to the contract and therefore not as good as the person who is being replaced. The reason for stating key people at tender stage is to ensure compatibility and experience; compatibility of the people working on the job is so important that clashes could jeopardise the smooth-running of the project. If the *Contractor* later changes the key people included at tender stage to less compatible people, this should be taken into account.

(1) The *Contractor*'s site manager has been assigned to another contract and is no longer able to work on the *Employer*'s contract.
(2) The *Contractor* decides who will replace him and gathers together his details for submission to the *Project Manager*.
(3) The *Contractor* submits the name of the proposed replacement person and his relevant qualifications and experience to the *Project Manager* for acceptance.
(4) The *Project Manager* reviews the qualifications and experience of the proposed replacement and compares them with the qualifications and experience of the current key person.
(5) Within the *period for reply*, the *Project Manager* notifies the *Contractor* whether he has accepted the proposed replacement or not. If he has not accepted the proposed replacement, he also includes his reason for his non-acceptance. The only reason[16] he may state is that the relevant qualifications and experience of the proposed replacement are not as good as those of the person who is to be replaced.
(6) If the reason for non-acceptance was not the reason stated in point (5) above, then the *Contractor* may notify a compensation event under

[13]Clause 14.4.
[14]Clause 14.4.
[15]Clause 24.1.
[16]Any other reason is a compensation event under clause 60.1(9) and may have time and money consequences.

clause 60.1(9). If the *Project Manager* did not reply within the *period for reply*, then the *Contractor* may notify a compensation event under clause 60.1(6).

(7) The *Contractor* hires the proposed replacement who has been accepted or proposes another replacement to the *Project Manager* for acceptance.

1.6.3.4 How the Project Manager *removes an employee*

This refers to a clause which sometimes causes difficulty in interpretation. The following is an attempt to demystify it and lend some explanation.

On reading the clause, *Employers* tend to think that the *Contractor* should not have one day to remove the employee, but that removal should take place immediately. In fact, the clause does not rule out the latter action occurring.

The clause allows the following:

- The *Project Manager* has the right to instruct the *Contractor* to remove any employee, which includes a Subcontractor's employees.
- The *Project Manager* has to provide reasons to the *Contractor* for his instruction to remove an employee, but there is no restriction on those reasons, there is no obligation to behave reasonably and the action does not result in a compensation event. It could be for health and safety reasons or because the employee is disruptive.
- The *Contractor* is required to arrange that, after one day, the employee has no further connection with the work included in this contract.[17]
- It might not be reasonable for the employee to cease work on the contract immediately, particularly if he has important papers or special knowledge about the contract, and so the *Contractor* has one day's grace to retrieve from the employee whatever he needs. This one day does not mean that the employee does not have to stop work immediately and leave the Site – that would be part of the *Project Manager*'s instruction – but only that he has one day to have no further connection with the work. If the *Project Manager* requires the removal to be immediate, then the *Contractor* is obliged to obey the *Project Manager*.

The *Project Manager* removes an employee as follows:

(1) The *Project Manager* gives written instruction to the *Contractor* that an employee of the *Contractor* is to be removed from Site immediately.

(2) The *Contractor* ensures that the employee leaves the Site and ensures that after one day the relevant employee has no further connection with the work included in the contract.

1.6.4 How to change the Working Areas

The Working Areas are those *working areas* that are identified by the *Contractor* in Contract Data part two. The *working areas* always include the Site and also include other areas that the *Contractor* may require to Provide the Works. The *working areas* would not include the head office of the *Contractor* or factories other than on-site fabrication shops. Only the *Contractor* may change the Working Areas since they are identified by him and are part of his Contract Data.

1.6.4.1 How the Contractor *changes the Working Areas*

(1) As a result of a large compensation event, the *Contractor* wishes to establish an on-site fabrication shop to produce small items of steelwork. The current Working Areas are too small and inconvenient to be on the site of the fabrication shop and the *Contractor* therefore wishes to add to the Working Areas. He designates an area for expansion.

(2) The *Contractor* submits to the *Project Manager* for acceptance a written proposal for making the addition to the Working Areas.

(3) The *Project Manager* considers whether the addition is necessary for Providing the Works and whether the area will be used for work for the contract.

(4) Within the *period for reply*, the *Project Manager* notifies his decision to the *Contractor* in writing. If he has not accepted the addition to the

[17] Clause 24.2.

Working Areas, the *Project Manager* must also include in the notification his reason for non-acceptance.

(5) If the reason for non-acceptance was not because the proposed addition is not necessary to Provide the Works or because the proposed area will be used for work not in the contract, then the *Contractor* may notify a compensation event under clause 60.1(9). If the *Project Manager* did not reply within the *period for reply*, then the *Contractor* may notify a compensation event under clause 60.1(6).

1.6.5 How to notify an early warning

Either the *Project Manager* or the *Contractor* could notify an early warning. An early warning[18] is a notification of a matter that could affect the price programme,[19] or performance of the *works* and it is required to be made as soon as the notifying party becomes aware of the matter.

In other words, if the *Contractor* thinks that it is possible that something will happen that will result in the cost of the project increasing **or** will result in Completion being later than the Completion Date **or** will affect the performance of the *works* (**or**, in ECC3, will delay meeting a Key Date) then he is obliged to notify the *Project Manager* immediately. It does not matter if the 'something' does not eventually materialise since the clause centres on the future, allowing the Parties to discuss[20] the matter and consider how best to deal with it.

An early warning is contractualised common sense and is an extension of acting in the spirit of mutual trust and cooperation. It tends to be far more productive and economical to sort out a problem before it occurs, rather than to wait until after the fact, when your options are reduced and the effects tend to be magnified.

There is no reply *per se* required to an early warning notification. Either the notifying party or the recipient of the notification may instruct the other to attend an early warning meeting[21] (called a risk reduction meeting in ECC3) if the matter is considered sufficiently urgent to require immediate attention. This is the only time that the *Contractor* may instruct the *Project Manager* to do something. Otherwise, early warnings may be discussed practically at a regular meeting, such as a weekly progress meeting.

There is a sanction[22] on the *Contractor* for not giving an early warning that he could have given. Of course, this does raise the question of whether the *Contractor* is an experienced contractor and whether he could have given the early warning. In general, the *Employer* would not have employed the *Contractor* if the latter were not experienced and, indeed, many forms of contract state as standard that the *Contractor* is an experienced contractor. Most contractors would not dispute that they are experienced. Whether they could have given an early warning may be subjective; however, documentation and observation will tend to show whether the *Contractor* had knowledge of the matter. It may simply boil down to the fact that an experienced *Contractor* should have had the knowledge and therefore could have given an early warning.

There is a slight overlap between early warnings and compensation events because compensation events, too, can be notified for events that have not yet happened.

In general, contractors should be aware of matters that could affect the project and that, if they occur, could reduce the *Project Manager*'s ability to manage the project effectively. It is therefore to the *Contractor*'s advantage to notify early warnings, especially since early warnings allow both Parties to project manage the project more effectively and economically.

[18] Clause 16.
[19] Completion Date in ECC2; Completion Date or Key Dates in ECC3.
[20] Clause 16.3.
[21] Clause 16.2.
[22] Clauses 61.5 and 63.4 (63.5 in ECC3).

The checklist for the *Contractor* is as follows:

(1) Is it possible that the matter could affect the project:
- prices,[23]
- programme (Completion Date in ECC2; Completion Date and Key Dates in ECC3),
- *works* once the *works* are in use?

(2) If so, give the *Project Manager* written notification **immediately** and instruct him to attend an early warning meeting if there are no other review meetings imminent at which the matter could be discussed and if the matter is urgent.

1.6.5.1 How the Contractor *notifies an early warning*

(1) The *Contractor* is notified by his supplier that a delivery of pipes for the next section of the pipeline he is building will be late. Although he has included some float in the programme, the *Contractor* considers that this late delivery could affect the project.

(2) That same day the *Contractor* gives the *Project Manager* written notification that he considers the late delivery could delay the Completion of the project. In the notification he also instructs the *Project Manager* to attend an early warning meeting the following day at 10 am.

1.6.5.2 How the Project Manager *notifies an early warning*

(1) The *Project Manager* has been notified by the *Employer*'s designer that the design of a central valve may change, depending on circumstances. The requirement for the valve is in two months time.

(2) That same day the *Project Manager* gives the *Contractor* written notification that he considers the changed design could increase the Prices, although it is unlikely to delay Completion.

(3) The *Project Manager* does not regard the matter with any urgency given the time-scales and so he does not instruct the *Contractor* to attend an early warning meeting/risk reduction meeting,[24] although he does add the matter to the agenda of the fortnightly progress meeting.

1.6.5.3 How to instruct attendance at an early warning meeting

Some *Project Managers* instigate early warning meetings/risk reduction meeting, separate to and in addition to progress or review meetings, although the early warning meetings/risk review meeting may take place immediately after the progress meeting. These meetings would suffice for matters that are of no immediate impact. However, there could be early warning matters that require immediate attention and problem solving. For these matters, the notifying party would instruct the other party to attend a meeting in the early warning notification. See the pro-formas for an illustration.

[23] This could depend on the main Option chosen since a delay might not affect an Option A, B, C or D contract Prices.

[24] In ECC3 the risk reduction meeting replaces the early warning meeting.

Pro-forma 6: EARLY WARNING NOTIFICATION BY THE *PROJECT MANAGER*

Project number:

Description:

Contract number:

Early warning number:

Date of this early warning notification:

To: (The *Contractor*)	
Address:	
Telephone:	
Fax:	
Attention:	

The matter notified is as follows:

The event could: (tick as appropriate)	Increase the total of the Prices	Delay Completion	Impair the performance of the *works* in use	Delay meeting a Key Date (ECC3 only)

The *Contractor* is instructed to attend an early warning meeting within . . . . day/hours of this notification.

Certified by the *Project Manager*:

Signature	Name	Date

Distribution:				

Pro-forma 7: EARLY WARNING NOTIFICATION BY THE *CONTRACTOR*

Project number:

Description:

Contract number:

Early warning number:

Date of this early warning notification:

To: (The *Project Manager*)	
Address:	
Telephone:	
Fax:	
Attention:	

The matter notified is as follows:

The event could: (tick as appropriate)	Increase the total of the Prices	Delay Completion	Impair the performance of the *works* in use	Delay meeting a Key Date (ECC3 only)

The *Project Manager* is instructed to attend an early warning meeting within . . . . day/hours of this notification.

Certified by the *Contractor*:

Signature   Name   Date

Distribution:				

1.6.6 How to notify an ambiguity or inconsistency

Either the *Contractor* or the *Project Manager* may notify the other immediately he becomes aware of an ambiguity or an inconsistency between the documents that make up the contract.[25] The *Project Manager* gives an instruction resolving the ambiguity or inconsistency.

The documents that are a part of the contract include the ECC *conditions of contract*, the Works Information, the Site Information, Contract Data parts one and two, the form of contract and any other document referred to in those documents. The most likely areas of inconsistencies are between the *conditions of contract* and the Works Information or between the documents that make up the Works Information. Since the Works Information yields the most likelihood of disputes and adjudications, it is important that this document is as sound as possible.

Where the instruction given by the *Project Manager* to resolve the ambiguity or inconsistency is a change to the Works Information, the contra proferentem rule is applied to the resulting compensation event:[26] if it is the Works Information provided by the *Employer* that is ambiguous or inconsistent, the interpretation is that which is most favourable to the *Contractor*; if it is the Works Information provided by the *Contractor* that is ambiguous or inconsistent, the interpretation is that which is most favourable to the *Employer*.

1.6.6.1 How the Contractor *notifies an ambiguity or inconsistency*

(1) The *Contractor* notices that one part of the Works Information states that water for volumetric tests will be provided by the *Employer*, while the testing part of the Works Information states that the *Contractor* will provide all materials and facilities for his own tests. He is unsure what is required of him.
(2) The *Contractor* gives the *Project Manager* immediate written notification of the inconsistency within the Works Information.
(3) The *Project Manager* considers the inconsistency and decides how to resolve it.
(4) The *Project Manager* gives the *Contractor* a written instruction changing the Works Information so that it is the *Employer* who will provide the water for the volumetric tests.
(5) Since a change to the Works Information is a compensation event, the *Project Manager* also notifies a compensation event to the *Contractor* and instructs the *Contractor* to submit quotations. The effect of the compensation event is assessed as if the Prices and the Completion Date were for the interpretation most favourable to the *Contractor*. In this instance the *Contractor* will have to consider the situation, and weigh up the facts as to what is most favourable to him, for example in ECC2 Prices, Completion Date and in ECC3 these plus the Key Dates. He will need to consider a range of factors, for example who pays to provide the water, does the *Employer* provide the water 'free of charge'?

1.6.6.2 How the Project Manager *notifies an ambiguity or inconsistency*

(1) The *Project Manager* notices an ambiguity or inconsistency in or between the documents which are part of the contract.
(2) The *Project Manager* gives the *Contractor* immediate written notification of the ambiguity or inconsistency in or between the documents which are part of the contract and how to resolve it.
(3) The *Project Manager* gives the *Contractor* a written instruction resolving the ambiguity or inconsistency.
(4) If the instruction resolving the ambiguity or inconsistency changes the Works Information it is a compensation event. If it is a compensation event the *Project Manager* will also notify a compensation event to the *Contractor* and instructs the *Contractor* to submit quotations. The effect of the compensation event is assessed in ECC2 as if the Prices and the Completion Date were for the interpretation most favourable to the *Contractor*. In ECC3 this interpretation will also include the Key Date.

[25] Clause 17.1.
[26] ECC2 clause 63.7; ECC3 clause 63.8.

1.6.7 How to notify an illegality or impossibility

Immediately the *Contractor* becomes aware that the Works Information requires him to do something that is illegal or impossible, he notifies the *Project Manager*.[27] If the *Project Manager* agrees that the Works Information requires the *Contractor* to do something that is illegal or impossible, he gives an instruction changing the Works Information. This change to the Works Information is also a compensation event and therefore the *Project Manager* also gives a notification of a compensation event.

1.6.7.1 How the Contractor *notifies that the Works Information requires him to do something that is illegal or impossible*

(1) The *Contractor* notices that the Works Information requires him to do something that is impossible and/or illegal.
(2) The *Contractor* gives the *Project Manager* immediate written notification that he considers that the Works Information requires him to do something that is impossible and/or illegal.
(3) The *Project Manager* considers the source of the alleged impossibility and determines whether the action is impossible. If he agrees that it is impossible, he decides how to resolve it.
(4) Within the *period for reply*, the *Project Manager* gives the *Contractor* either:
- notification that he agrees that the Works Information requires the *Contractor* to do something that is impossible or illegal and a written instruction changing the Works Information appropriately to resolve the impossibility or
- notification that he does not agree that the Works Information requires the *Contractor* to do something that is impossible and possibly an explanation of why the action is not impossible.

(5) If the *Project Manager* has changed the Works Information, he also notifies a compensation event to the *Contractor* and instructs the *Contractor* to submit quotations.
(6) If the *Contractor* is unhappy with the *Project Manager*'s decision that there is no impossibility, he may take the matter to adjudication.

1.6.8 How to deal with prevention (ECC3 only)

If an event occurs which neither Party could prevent and which an experienced contractor thought was unlikely to occur and the event stops the *Contractor* from completing the *works* or from completing the *works* by the date on the Accepted Programme, then the *Project Manager* gives an instruction to the *Contractor* stating how to deal with the event.[28]

1.6.9 How the *Contractor* submits his design

The *Contractor* designs those parts of the *works* that the Works Information states he is to design.[29] The Works Information should therefore include a statement telling the *Contractor* what he is to design, for example the whole of the *works*, the parts of the *works* not designed by the *Employer*, only the connections, or no part of the *works*. If both the *Employer* and the *Contractor* are involved in design work, the interfaces between the two should also be clearly identified and stated.

The Works Information should also state how and when the *Contractor* is required to submit the particulars of his design.[30] The *Contractor* may not proceed with the relevant work until the *Project Manager* has accepted the design[31] and the *Contractor* should be careful to incorporate this time period into his programme.

If the design can be assessed fully in parts, without reference to other aspects of the design, then the *Contractor* may submit his design in parts to the *Project Manager* for acceptance.[32]

The *Employer* may use and copy the *Contractor*'s design for any purpose stated in the Works Information.[33] Note that there is no standard clause in the ECC relating to ownership of design.

[27] ECC2 clause 19.1; ECC3 clause 18.1.
[28] ECC3 clause 19.1.
[29] Clause 21.1.
[30] Clause 21.2.
[31] Clause 21.2.
[32] Clause 21.3.
[33] Clause 22.1.

The *Contractor* may also be required to design items of Equipment (items provided by the *Contractor* to Provide the Works but which are not included in the *works*, e.g. temporary sheet piling, crash decks, shoring) and the *Project Manager* may instruct the *Contractor* to submit the particulars of his design of Equipment to him for acceptance.[34]

1.6.9.1 How the Contractor *submits his design to the* Project Manager *for acceptance*

The submission of the *Contractor*'s design is dependent on the Works Information. A general submission is related below.

(1) The *Contractor* designs those parts of the *works* that he is required to design in accordance with the Works Information.
(2) The *Contractor* submits the particulars of his design to the *Project Manager* for acceptance.
(3) The *Contractor* may not commence the work described on the submitted design.
(4) The *Project Manager* considers the design and possibly its relation to the Works Information submitted by the *Contractor* at tender stage and incorporated into the contract.
(5) Within the *period for reply*, the *Project Manager* notifies the *Contractor* of his acceptance or otherwise of the design. The *Project Manager* may only give two reasons for not accepting the design:[35] that it does not comply with the Works Information (either by the *Employer* or by the *Contractor*); or that it does not comply with the applicable law.
(6) If the *Project Manager* has accepted the *Contractor*'s design, the *Contractor* may commence the work.
(7) If the *Project Manager*'s non-acceptance of the design was for a reason other than that the design did not comply with the Works Information or the applicable law, the *Contractor* may notify a compensation event.
(8) If the *Project Manager* did not reply within the *period for reply*, then the *Contractor* may notify a compensation event.
(9) If the *Project Manager*'s non-acceptance of the design was for a reason included in the contract, but the *Contractor* does not agree, he may notify a dispute in accordance with the contract. The exact procedure will depend on whether Y(UK)2 (ECC2) has been included or Option W1 or Option W2 (ECC3).

1.6.10 How to subcontract

A Subcontractor is a defined term[36] and in ECC2 includes only those bodies who provide parts of the *works* or Plant and Materials[37] for the *works*.

Because the wording in ECC2 was considered to be a little ambiguous, in ECC3 the defined term has been redrafted to create greater clarity:

> 'A Subcontractor is a person or organisation who has a contract with the *Contractor* to
>
> - construct or install part of the *works*
> - provide a service necessary to Provide the Works or
> - supply Plant and Materials which the person or organisation has wholly or partly designed specifically for the *works*.'

The Works Information could give an indication of what the *Contractor* may or may not subcontract. For example, for a mechanical and piping contract, the welding of the pipes could be an acceptable subcontract, but the manufacture of the pipes might not be.

There is no nominated subcontracting in the ECC. If employers want a particular contractor to do the painting or the scaffolding, for example, then the employer should contract separately with this contractor and allow the *Contractor* for the *works* to interface with the scaffolding or painting contractor as required.

[34] Clause 23.1.
[35] Without invoking a compensation event.
[36] ECC2 clause 11.2(9); ECC3 clause 11.2(17).
[37] ECC2 clause 11.2(10); ECC3 clause 11.2(12).

In traditional tenders, the *Employer* may be used to requesting a list of Subcontractors to whom the *Contractor* intends to subcontract parts of the *works*. The *Contractor* then includes in his tender the list of Subcontractors that he proposes to use. This request is not ultimately necessary in the ECC, but could be used to expedite matters or to maintain some control over the procedure. In addition, the tender documentation could include a procedure for submitting information about the Subcontractors so that the *Contractor* is able to provide all the information that the *Project Manager* requires in a timely manner. The *Project Manager* would still be obliged to reply to the *Contractor* regarding the acceptance or otherwise of the proposed Subcontractors.

The *Contractor* may not appoint a Subcontractor until the *Project Manager* has accepted him. If the *Contractor* does appoint a Subcontractor for substantial work prior to the *Project Manager*'s acceptance, the *Employer* may terminate the contract.[38]

The *Contractor* is also required to submit the proposed conditions of subcontract to the *Project Manager* for acceptance unless the ECS[39] or PSC[40] is to be used in ECC2 or an NEC contract is proposed in ECC3 or unless the *Project Manager* has agreed that such a submission is not required.[41] The *Contractor* does not appoint the Subcontractor on those conditions until the *Project Manager* has accepted them.

For options C, D, E and F, the *Contractor* is also required to submit the Contract Data for the subcontract to the *Project Manager* for acceptance if the ECS or PSC (or NEC in ECC3) is to be used and if the *Project Manager* instructs the *Contractor* to make such a submission.[42]

1.6.10.1 How the Contractor *proposes a Subcontractor*

The ECC refers to the notification regarding Subcontractors as being between the *Project Manager* and the *Contractor*. This would indeed be the case if additional Subcontractors were proposed during the period of the contract due to a compensation event or due to the *Contractor*'s way of working, especially during a design-and-build contract. In some cases, however, depending on the *works* and the time-scales involved before the *Contractor* is required to engage his Subcontractors, the acceptance of Subcontractors and the subcontract *conditions of contract* will have to take place prior to the Contract Date in order to allow the *Contractor* sufficient time to appoint the Subcontractors before they have to start work. If this is the case, then it is debatable whether the restrictions and procedures of the ECC *conditions of contract* still apply. It is advised, however, that the procedure is followed as closely as possibly, even though the *period for reply* and the reasons for non-acceptance do not actually apply until the contract is executed.

The progress of the procedure for appointing Subcontractors is not clear from the ECC *conditions of contract*, but is elucidated in the ECC2 flow charts. From the ECC flow charts, it appears that the *Contractor* has to propose the names of Subcontractors first, then await further instructions before proposing the conditions of subcontract or Contract Data. An interesting side point here is the effect that the acceptance/non-acceptance may have on the *Contractor's subcontracted fee percentage*.[43]

It is likely, however, that the submission of proposed *conditions of contract* and proposed Contract Data[44] would depend largely on the procedures discussed by the *Project Manager* at a start-up meeting or included in the tender documentation. Given that the *Contractor* may not engage the Subcontractor until he and his conditions of subcontract have been accepted, the timing may

[38] ECC2 clause 95.2; ECC3 clause 91.2.
[39] Engineering and Construction Subcontract (NEC).
[40] Professional Services Contract (NEC).
[41] Clause 26.3.
[42] Clause 26.4.
[43] See Chapter 2 of Book 4 and Appendix 2 of Chapter 2 of Book 4 for further discussion on this point.
[44] For Options C, D, E and F only.

impinge on the *Contractor*'s programme and this might be something that the *Contractor* wishes to discuss with the *Project Manager* before the contract is executed.

The *Project Manager* could have stated that he wants to see only the *conditions of contract* and Contract Data for Subcontractors whose work comprises a substantial part of the *works*, for example

- the designer of a bridge in a design-and-build contract or
- the manufacturer of steelwork in a structure.

He could also waive his right to see any *conditions of contract* by agreeing that no submission is required. The *Project Manager*'s wanting to see proposed *conditions of contract* and Contract Data could depend on his commitment to the tenets of the ECC and his desire to see the project as a whole working to the same philosophies, rather than just the main contract. It could also depend on the importance of the *Contractor* having back-to-back contracts with his Subcontractors.

The following description assumes that the instructions to tenderers or the tender documentation has requested a list of Subcontractors and has also described procedures for the submission of *conditions of contract* and Contract Data. The procedures may also appear in the Works Information for a design-and-build contract or where Subcontractors are appointed after the Contract Date.

The *Employer*'s representative during the tender evaluation period will notify the tenderer whether the Subcontractors proposed by him in his tender are acceptable or not. In this same notification, the *Employer*'s representative may instruct the tenderer to submit proposed *conditions of contract* for some or all of the proposed subcontracts and may also instruct the tenderer to submit proposed Contract Data for ECS or PSC subcontracts (or NEC in the ECC3).

(1) The instructions to tenderers include an instruction for the tenderer to propose a list of Subcontractors that he intends to use and describes the subsequent procedure to be followed.

(2) The tenderer compiles a list of subcontractors to whom he intends to subcontract work, having regard to the elements of the work that he is allowed to subcontract in accordance with the Works Information. He submits the list with his tender.

(3) The *Employer*'s representative will during the tender assessment period consider the names of the Subcontractors and decide whether or not to accept them. He will notify the tenderer of his acceptance or otherwise of the proposed Subcontractors. The *Employer*'s representative's only reason[45] for not accepting the Subcontractor[46] is that he will not allow the *Contractor* to Provide the Works.[47]

(4) With his notification of acceptance, the *Employer*'s representative also informs the tenderer whether he needs to see the conditions of subcontract and whether he wishes to see the Contract Data for any proposed ECS and PSC (NEC in the ECC3) subcontracts.[48]

(5) The tenderer may not appoint the proposed Subcontractors until the *Employer*'s representative has accepted the proposed conditions of subcontract and/or the Contract Data.

(6) If the *conditions of contract* are not the ECS or PSC (NEC in the ECC3) and if the *Employer*'s representative has instructed the tenderer to submit the proposed conditions of subcontract, then the tenderer must

[45]Without invoking a compensation event, although since the contract has not yet been executed, this contractual procedure may not be used.
[46]Bearing in mind that the contract has not yet been executed and so the *Employer* is not restricted to the reasons included in the *conditions of contract*.
[47]Given the definition of Providing the Works (ECC2 clause 11.2(4), ECC3 clause 11.2(13)), experience of a disruptive subcontractor that will impede the *Contractor*'s progress ('completing the *works* in accordance with this contract', that is, by the Completion Date) could fall under the gambit of the reason allowed.
[48]Where the main contract is Option C, D, E or F only.

submit the proposed conditions of subcontract to the *Employer*'s representative for acceptance within the *period for reply*.

(7) The tenderer must submit the proposed Contract Data for each subcontract to the *Employer*'s representative for acceptance within the *period for reply*:
- if the main Option for the main *works* contract is C, D, E or F and
- if the conditions of subcontract for each accepted Subcontractor are the ECS or the PSC (NEC in the ECC3) and
- if the *Project Manager* has so instructed.

(8) Within the *period for reply*, the *Employer*'s representative must reply to the tenderer accepting the proposed conditions of subcontract or giving his reasons for not accepting them. His reasons for non-acceptance can only be that the proposed conditions of subcontract will not allow the *Contractor* to Provide the Works or that the proposed conditions of subcontract do not include a statement that the Parties to the subcontract shall act in a spirit of mutual trust and cooperation.

Within the same reply, the *Employer*'s representative must accept the proposed Contract Data for each subcontract of the ECS or PSC (NEC in the ECC3) or provide his reasons for not accepting them. His only reason for non-acceptance is that they will not allow the *Contractor* to Provide the Works.

(9) If the *Employer*'s representative has not accepted the proposed *conditions of contract* and/or Contract Data, the tenderer must resubmit for acceptance of them within the *period for reply*.

(10) If the *Employer*'s representative has accepted the proposed *conditions of contract* and/or Contract Data, the tenderer may appoint the Subcontractors on the accepted conditions of subcontract and/or Contract Data.

1.6.11 Completion Completion is a defined term and it is described in the Works Information.[49] The ECC does not cater for mechanical completion, practical completion or substantial completion and any inclusion of such terms in the Works Information should be clearly described. Completion occurs when the *Contractor* has done everything the Works Information requires him to do by the Completion Date.[50] This could include as-built drawings or operational manuals, or you could ask to receive these things within a specified period after Completion. In the latter case, there is little incentive for the *Contractor* to produce the described items since he would have been paid at Completion prior to submitting the documents.

Completion can only take place when the *Contractor* has corrected notified Defects, which would have prevented the *Employer* from using the *works*[51] and Others from doing their work (ECC3). This correlates with clause 43.1[52] where Defects notified before Completion are only corrected after Completion, except obviously those that prevent the *Employer* from using the *works* (or Others from doing their work in ECC3) otherwise Completion would not be achieved in the first place.

It is most important to describe Completion in the Works Information in objective terms so that the *Project Manager* can determine whether or not Completion has been achieved.[53] Difficulties may arise where the *Employer* fails to describe Completion in the Works Information.[54] *Employers* often omit to describe Completion because they are still thinking in terms of practical or substantial or

[49] ECC2 clause 11.2(13), ECC3 clause 11.2(2).
[50] ECC2 clause 11.2(13), ECC3 clause 11.2(2).
[51] ECC2 clause 11.2(13), ECC3 clause 11.2(2).
[52] ECC3 clause 43.2.
[53] Clause 30.2.
[54] Note that ECC3 contains a default to the definition of Completion in clause 11.2(2): where the work which the *Contractor* is to do by the Completion Date is not stated in the Works Information, Completion is when the *Contractor* has done all the work necessary for the *Employer* to use the *works* and for Others to do their work.

mechanical completion – terms that everybody understands (or do they?). Because practical or substantial or mechanical completion is not included in the Works Information, nor described, the problem remains.

Completion is a status and it is a separate concept from the Completion Date. Completion could take place on, before or after the Completion Date. The *Project Manager* decides the date of Completion in accordance with the definition included in the Works Information.

1.6.11.1 How the Contractor *reaches Completion*

The *Contractor* reaches Completion when the status of the *works* matches the description of Completion in the Works Information. In the case of Completion of *sections*, there would be the same number of different descriptions of Completion as there are *sections* of the *works*.

1.6.11.2 How the Contractor *notifies Completion*

The *Contractor* is not required to notify Completion in the ECC. It could be part of the *Contractor*'s internal procedures as a result of working with traditional contracts, but this notification is not required under the ECC, does not require a reply from the *Project Manager* and acquires no status under the ECC.

1.6.11.3 How the Project Manager *notifies Completion*

(1) The *Project Manager* examines the *works* and assesses whether the *works* are complete or not based on the definition of Completion and the description of Completion in the Works Information. Once he decides that Completion has been achieved, he makes a note of that date.
(2) The *Project Manager* will no doubt be in constant communication with the *Contractor* on the Site and will possibly verbally inform the *Contractor* that he considers that Completion has been achieved. He may also inform the *Contractor* that he (the *Project Manager*) expects a final statement of account from the *Contractor* in due course to allow the assessment at Completion to take place.
(3) The *Project Manager* certifies Completion within one week of Completion and issues his certificate to the *Contractor* and the *Employer*. The certificate includes the date of Completion, the Completion Date and the *defects date*, which is a period of time (such as 52 weeks) after Completion (rather than 52 weeks after the Completion Date).
(4) The *Project Manager* assesses the amount due. Payment for the assessment taking place on Completion should include the return of half of any retention that has been retained by the *Employer*.
(5) Take over takes place not more than two weeks after Completion.

1.6.12 Take over

Take over takes place within two weeks after Completion.[55]

The principal reason for identifying take over is to mark the point where loss of or damage to the *works* becomes an *Employer*'s risk.[56] Take over may trigger a compensation event where it happens before both Completion or the Completion Date[57] unless the take over falls within the parameters described in clause 35.3 (ECC3 clause 35.2) and included in the Works Information. The *Employer* does not have to take over the *works* before the Completion Date if the optional statement in the Contract Data was included.[58]

Take over is also important because it is the primary reason for an *Employer* to choose Option L sectional completion (ECC3 Option X5). If the *Employer* wants to take over parts of the *works* as they are completed, then Option L/Option X5 should be chosen as part of the contract strategy, where the *completion date* for different parts of the *works* may be described. Take over for each part would take place within two weeks of each Completion.

It should be noted that there is no allowance within the ECC to have different *defects dates* for different *completion dates*. The *defects date* is a period of

[55]ECC2 clause 35.2; ECC3 clause 35.3.
[56]Clause 80.1.
[57]Clause 60.1(15).
[58]ECC2 clause 35.2; ECC3 clause 35.1.

time after Completion of **the whole of the *works***. If, therefore, the contract is of long duration with sectional completion and multiple take over, the *defects date* is still triggered by the last Completion – that of the whole of the *works*. Sections of the *works* completed earlier would therefore be subjected to a longer period during which the *Contractor* has to correct Defects and so on. To amend this situation, the *Employer* could alter the relevant statement in the Contract Data to give different *defects dates* for different *sections* of the *works*. This could increase the administrative burden and may not work if the functioning of the *works* as a whole is dependent on the adequate functioning of all of its parts.

1.6.12.1 How the Project Manager *certifies take over*

(1) The *Project Manager* has decided the date of Completion and has certified Completion within one week of Completion.
(2) The *Employer* takes over the *works* not more than two weeks after Completion. This may refer to Completion of a *section* or Completion of the whole of the *works*.
(3) Possession/access[59] of each part of the Site taken over returns to the *Employer* at take over.
(4) The *Project Manager* certifies take over not more than one week after the date of take over. See Fig. 1.2 below.

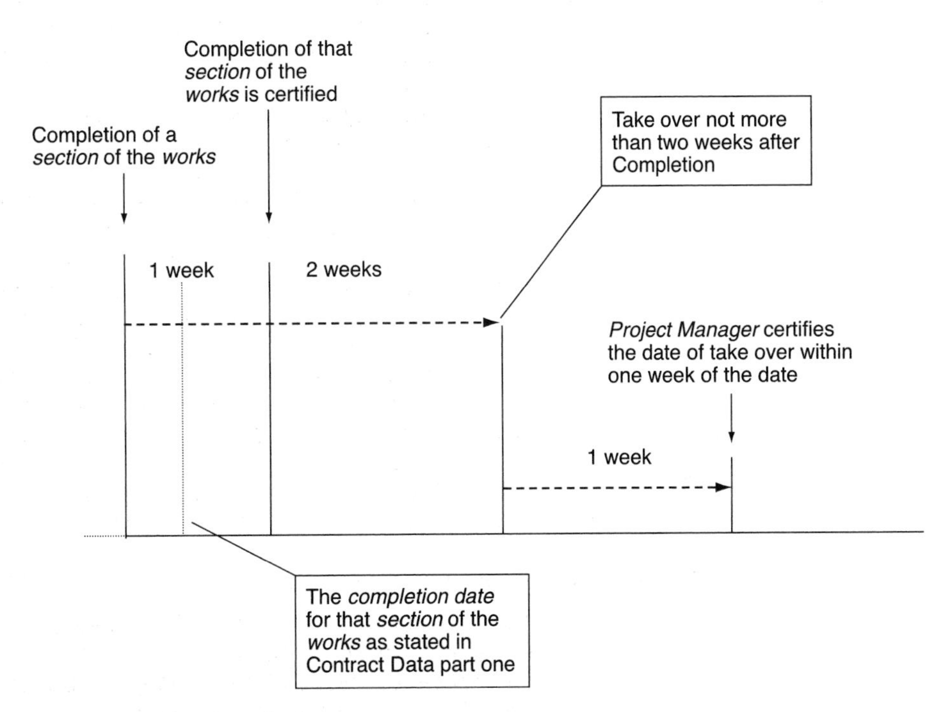

Fig. 1.2. Certification of take over

[59] ECC2 possession = ECC3 access.

Pro-forma 8: COMPLETION CERTIFICATE

Project number:

Description:

Contract number:

To: (The *Contractor*)		To: (The *Employer*)	
Address:		Address:	
Telephone:		Telephone:	
Fax:		Fax:	
Attention:		Attention:	

	Day	**Month**	**Year**
The Completion Date is:			
Completion achieved on:			
Date of this Completion certificate (within one week of Completion):			
The *defects date* is (period after Completion):			
The Defects on the attached schedule are to be corrected within the *defect correction period* which ends on:			

Half the retention amount held will be released in the next payment certificate:

The *Employer* takes over the *works* not more than two weeks after Completion:

***Works* checked by *Supervisor*:**

Signature Name Date

Certified by *Project Manager*:

Signature Name Date

Distribution:				

Pro-forma 9: SECTIONAL COMPLETION CERTIFICATE (ECC2 OPTION L; ECC3 OPTION X5)

Project number:

Description:

Contract number:

Section of the *works*:

To: (The *Contractor*)		To: (The *Employer*)	
Address:		Address:	
Telephone:		Telephone:	
Fax:		Fax:	
Attention:		Attention:	

	Day	**Month**	**Year**
The sectional Completion Date is:			
Sectional Completion achieved on:			
Date of this sectional Completion certificate (within one week of Completion):			
The *defects date* is (period after Completion of the whole of the *works*):			
The Defects on the attached schedule are to be corrected within the *defect correction period* which ends on:			

The Employer takes over the *works* not more than two weeks after Completion:

***Works* checked by *Supervisor*:**

Signature Name Date

Certified by *Project Manager*:

Signature Name Date

Distribution:				

Pro-forma 10: TAKE OVER CERTIFICATE

Project number:

Description:

Contract number:

To: (The *Contractor*)		To: (The *Employer*)	
Address:		Address:	
Telephone:		Telephone:	
Fax:		Fax:	
Attention:		Attention:	

Events	**Day**	**Month**	**Year**
Completion was achieved on:			
The Completion Date is:			
The take over date is (2 weeks after Completion):			
The date of this take over certificate is (within one week of take over):			

Certified by *Project Manager*:

Signature ____________ Name ____________ Date ____________

Distribution:				

Pro-forma 11: TAKE OVER CERTIFICATE FOR A *SECTION* OF THE *WORKS*

Project number:

Description:

Contract number:

Section of the *works*:

To: (The *Contractor*)		To: (The *Employer*)	
Address:		Address:	
Telephone:		Telephone:	
Fax:		Fax:	
Attention:		Attention:	

Events	**Day**	**Month**	**Year**
Sectional Completion was achieved on:			
The *section* Completion Date is:			
The take over date is (2 weeks after Completion):			
The date of this take over certificate is (within one week after take over):			

Certified by *Project Manager*:

Signature   Name   Date

Distribution:				

1.6.13 Programme

1.6.13.1 How the Contractor *submits his programme*

The *Contractor* is required to submit a first programme as chosen by the *Employer* either:

- at tender stage or
- a period of time after the contract is executed (the Contract Date).

The *Contractor* is also required to submit revised programmes:

- when instructed to do so by the *Project Manager*,
- every period of time as stated in the Contract Data,
- when he chooses to do so and
- with a quotation for a compensation event, where the programme has been affected (this is referred to as alterations to the Accepted Programme in ECC3).

1.6.13.2 Where the tender documents require a programme at tender stage

(1) The tender documents include a statement about the programme in Contract Data part two requiring a programme with the tender.

(2) The tenderer submits with his tender the required programme showing all the required information including resource and method statements.
(3) Before the execution of the contract, the *Project Manager* informs the *Contractor* whether he has accepted the programme or not. There are four reasons why he may not accept the programme:
(4) If the *Project Manager* has accepted the *Contractor*'s programme, it becomes the Accepted Programme.
(5) If the *Project Manager*'s non-acceptance of the programme was for a reason other than the four included in clause 31.3, the *Contractor* may notify a compensation event.
(6) If the *Project Manager*'s non-acceptance of the programme was for a reason included in the contract, but the *Contractor* does not agree, he may notify a dispute in accordance with the contract (the exact procedure will depend on whether Y(UK)2 has been included for ECC2 and Options W1 and W2 for ECC3).
(7) If the *Project Manager* did not accept the programme, the *Contractor* resubmits the programme within the *period for reply*, taking into account the reasons for non-acceptance.

1.6.13.3 Where the tender documents require a programme two weeks after the Contract Date

(1) The invitation to tender includes a statement about the programme in Contract Data part one requiring a programme two weeks after the Contract Date.
(2) The *Contractor* submits his programme two weeks after the Contract Date, showing all the required information including resource and method statements.
(3) Within two weeks of this submission, the *Project Manager* informs the *Contractor* whether he has accepted the programme or not. There are four reasons why he may not accept the programme:
(4) If the *Project Manager* has accepted the *Contractor*'s programme, it becomes the Accepted Programme.
(5) If the *Project Manager*'s non-acceptance of the programme was for a reason other than the four included in clause 31.3, the *Contractor* may notify a compensation event.
(6) If the *Project Manager* did not reply within the *period for reply* (say two weeks), then the *Contractor* may notify a compensation event.
(7) If the *Project Manager*'s non-acceptance of the programme was for a reason included in the contract, but the *Contractor* does not agree, he may notify a dispute in accordance with the contract (the exact procedure will depend on whether Y(UK)2 has been included in ECC2 and Options W1 and W2 in ECC3).
(8) If the *Project Manager* did not accept the programme, the *Contractor* resubmits the programme within the *period for reply*, taking into account the reasons for non-acceptance.
(9) If the first programme was not submitted within two weeks of the Contract Date and has not been submitted by the time of the first assessment date, the *Project Manager* may deduct 25% of the Price for Work Done to Date until a programme is submitted.
(10) If there is no Accepted Programme at the time that assessment of a compensation event is required, then the *Project Manager* may use his own assessment of the programme.
(11) If the *Project Manager* has not accepted the *Contractor*'s latest programme for a reason in the contract, then he may assess a compensation event that arises in the meantime.

1.6.13.4 Submitting a revised programme as stated in the Contract Data

(1) If the interval stated in the Contract Data has passed (say eight weeks) since the last programme submission, then the *Contractor* revises his programme including the additional information required as described in clause 32.1.
(2) Within two weeks of this submission, the *Project Manager* informs the *Contractor* whether he has accepted the programme or not. There are four reasons why he may not accept the programme:

(3) If the *Project Manager* accepts this revised programme, it becomes the Accepted Programme.
(4) If the *Project Manager* does not accept it, the previous programme that was accepted remains the Accepted Programme until a revised programme is accepted.
(5) If the *Project Manager*'s non-acceptance of the programme was for a reason other than the four included in clause 31.3, the *Contractor* may notify a compensation event.
(6) If the *Project Manager* did not reply within the *period for reply* (say two weeks), then the *Contractor* may notify a compensation event.
(7) If the *Project Manager*'s non-acceptance of the programme was for a reason included in the contract, but the *Contractor* does not agree, he may notify a dispute in accordance with the contract (the exact procedure will depend on whether Y(UK)2 has been included in ECC2 and Options W1 and W2 in ECC3).
(8) If the *Project Manager* did not accept the programme, the *Contractor* resubmits the programme within the *period for reply*, taking into account the reasons for non-acceptance.

1.6.13.5 Submitting a revised programme as instructed by the Project Manager

(1) If the *Project Manager* instructs the *Contractor* to submit a programme, he is required to submit a revised programme within the *period for reply*.
(2) Within two weeks of this submission, the *Project Manager* informs the *Contractor* whether he has accepted the programme or not. There are four reasons why he may not accept the programme:
(3) If the *Project Manager* accepts this revised programme, it becomes the Accepted Programme.
(4) If the *Project Manager* does not accept it, the previous programme that was accepted remains the Accepted Programme until a revised programme is accepted.
(5) If the *Project Manager*'s non-acceptance of the programme was for a reason other than the four included in clause 31.3, the *Contractor* may notify a compensation event.
(6) If the *Project Manager* did not reply within the *period for reply* (say two weeks), then the *Contractor* may notify a compensation event.
(7) If the *Project Manager*'s non-acceptance of the programme was for a reason included in the contract, but the *Contractor* does not agree, he may notify a dispute in accordance with the contract (the exact procedure will depend on whether Y(UK)2 has been included in ECC2 and Options W1 and W2 in ECC3).
(8) If the *Project Manager* did not accept the programme, the *Contractor* resubmits the programme within the *period for reply*, taking into account the reasons for non-acceptance.

1.6.13.6 Submitting a revised programme because the Contractor *chooses to*

(1) The *Contractor* may choose to submit a programme, for example if he has changed a planned sequence of work or statement[60] of how he plans to carry out the work for an operation.
(2) Within two weeks of this submission, the *Project Manager* informs the *Contractor* whether he has accepted the programme or not. There are four reasons why he may not accept the programme.
(3) If the *Project Manager* accepts this revised programme, it becomes the Accepted Programme.
(4) If the *Project Manager* does not accept it, the previous programme that was accepted remains the Accepted Programme until a revised programme is accepted.
(5) If the *Project Manager*'s non-acceptance of the programme was for a reason other than the four included in clause 31.3, the *Contractor* may notify a compensation event.
(6) If the *Project Manager* did not reply within the *period for reply* (say two weeks), then the *Contractor* may notify a compensation event.

[60] ECC2 refers to method statements.

(7) If the *Project Manager*'s non-acceptance of the programme was for a reason included in the contract, but the *Contractor* does not agree, he may notify a dispute in accordance with the contract (the exact procedure will depend on whether Y(UK)2 has been included in ECC2 and Options W1 and W2 in ECC3).

(8) If the *Project Manager* did not accept the programme, the *Contractor* resubmits the programme within the *period for reply*, taking into account the reasons for non-acceptance.

1.6.13.7 Submitting a revised programme[61] with a compensation event quotation

(1) The *Contractor* may be required to submit a programme as part of a compensation event quotation showing the effects of the compensation event if the programme for the remaining work is affected by the compensation event. The effect of the compensation event could be a change in resources, a change in planned sequence of work or statement[62] of how he plans to carry out the work for an operation, or a delay to the Completion Date as measured by the length of time that planned Completion is later than planned Completion as shown on the Accepted Programme (or the effect on a Key Date in ECC3).

(2) The *Project Manager* assesses the revised programme[63] together with the rest of the quotation and replies within two weeks of the submission of the quotation. His reply may be one of four notifications.

(3) If the *Contractor* was required to submit a revised programme with his quotation but he did not do so, the *Project Manager* may assess the compensation event.

1.6.13.8 How the Project Manager *replies to the Contractor's programme*

(1) The *Project Manager* must reply to a programme submission within two weeks of the submission. If he replies later than two weeks after the programme submission his reply is probably still valid, since the *Contractor* may notify a compensation event regarding the lateness of the reply. If he does not reply at all, acceptance of the programme will probably be deemed to have occurred.[64]

(2) If the *Project Manager* accepts the programme, it becomes the Accepted Programme and replaces any previous Accepted Programmes.

(3) If the *Project Manager* does not accept the programme, the *Contractor* submits a further revised programme within the *period for reply*. The *Project Manager* has two weeks to reply to this programme.

(4) If the *Project Manager*'s reason for non-acceptance of the programme is not one of the following, then the *Contractor* may notify a compensation event:
- the *Contractor*'s plans which it shows are not practicable,
- it does not show the information which this contract requires,
- it does not represent the *Contractor*'s plans realistically,
- it does not comply with the Works Information.

1.6.14 Progress

1.6.14.1 How the Employer *gives possession (ECC2) (ECC3 access) and use of the site*

Possession (access) does not mean that the *Contractor* possesses the Site but that he has been given licence to occupy the site up to the date of completion, to enter the land and carry out the work.[65]

The *possession date* (*access date*) is different from the *starting date*. The *starting date* is the date when the *Contractor* starts the work that he is required to do before he comes on to the Site. The *possession date* (*access date*) is when he may start work on Site.[66] Mobilisation can take place before the *possession date* (*access date*) but work on the Site may only start on the *possession date* (*access date*). Both the *starting date* and the *possession dates* (*access date*) are identified in the Contract Data.

[61] ECC3 refers to alterations to the Accepted Programme.
[62] ECC2 refers to method statements.
[63] Alterations to the Accepted Programme in ECC3.
[64] The ECC is silent on a non-reply since it assumes that the *Project Manager* will do what he is obliged to do.
[65] B. Eggleston, *The New Engineering Contract: a commentary*, Blackwell Science, Oxford, 1996, p. 144.
[66] Clause 30.1.

The *Contractor* also identifies on his programme the date by which he requires possession (access), and this date may be later than the *possession date* (*access date*) included in the Contract Data (but it may not be earlier) while still achieving Completion on or before the Completion Date. The *Employer* must give possession (access) of each part of the Site to the *Contractor* on or before the later of its *possession date* (*access date*) and the date for possession/access shown on the Accepted Programme[67] and it is a compensation event if the *Employer* does not do so.[68]

This is explained as follows. In Contract Data part one, the *Employer* will have included *possession dates* (*access dates*) for each part of the Site where the Site is divided, or for the whole of the Site where it is not divided. In his first programme submitted to the *Project Manager* that has been accepted, the *Contractor* will have included dates when he requires possession/access of the Site or its relevant parts. If the *possession date* (*access date*) stated by the *Employer* and the date for possession/access required by the *Contractor* are different, it is the later of the two that provides the latest date by which the *Employer* may give the *Contractor* possession/access.

In ECC2 *possession* of each part of the Site returns to the *Employer* at take-over[69] and *possession* of the whole of the Site returns to the *Employer* when the *Project Manager* certifies termination.[70] There are no direct equivalent words in ECC3. However, ECC3 introduces the new identified term *access* to replace the ECC2 phrase *possession* to signify that the *Employer* allows *access* to the Site rather than giving the Contractor *possession* of the Site and as such any equivalent words are not relevant.

While the *Contractor* has possession/access of the Site, the *Employer* gives the *Contractor* access to and use of the Site.[71] Access is also given to the *Contractor* after take over if access is needed in order to correct a Defect.[72]

1.6.14.2 How the Project Manager *stops work*

The *Project Manager* may instruct the *Contractor* to stop or not to start any work and may later instruct him that he may restart or start it.[73] An instruction to stop or not to start any work is a compensation event.[74] If such an instruction has been given and an instruction to restart or start the work has not been given within 13 weeks of the original instruction to stop or not to start, then either Party may terminate the contract.[75]

(1) The *Project Manager* instructs the *Contractor* to stop work. He includes in his instruction notification that the reason for the instruction is a dangerous violation of the health and safety regulations as stated in the Works Information.
(2) The *Contractor* stops work immediately.
(3) Within one week of the instruction to stop work, the *Contractor* has rectified the violation.
(4) The *Project Manager* instructs the *Contractor* to restart the work.
(5) Within two weeks of the instruction to stop work, the *Contractor* notifies a compensation event in accordance with clause 60.1(4).
(6) Within one week of the *Contractor*'s notification, the *Project Manager* notifies the *Contractor* that the compensation event arose from the *Contractor*'s default and in ECC2 that the Prices and the Completion Date would not be changed. In ECC3 the Key Dates are also not changed.
(7) If the *Project Manager* had not been able to instruct a restart to the work within 13 weeks of the instruction to stop work due to the

[67] Clause 33.1.
[68] Clause 60.1(2).
[69] ECC2 clause 35.1.
[70] ECC2 clause 35.1 and clause 96.2; ECC3 clause 92.2.
[71] ECC2 clause 33.2; ECC3 clause 33.1.
[72] ECC2 clause 43.3; ECC3 clause 43.4.
[73] Clause 34.1.
[74] Clause 60.1(4).
[75] ECC2 clause 95.6; ECC3 clause 91.6.

Contractor's not rectifying the default, then the *Employer* would have grounds for termination of the contract.[76]

1.6.14.3 How the Project Manager *notifies acceleration*

Acceleration is covered in clause 36 and there are also some clauses in the main Options[77] denoting the different ways of implementing acceleration for the different main Options.

Although acceleration may mean different things in different contracts, it has a specific meaning under the ECC: that of shortening the time allowed for completion by bringing the Completion Date forward in time.

The *Project Manager* may **not** instruct the *Contractor* to accelerate. He may only instruct the *Contractor* to submit a quotation to do so[78]. The *Contractor* may either submit a quotation or give his reasons for not submitting a quotation.[79] In other words, the *Contractor* may choose whether to accelerate or not. He is not obliged to accelerate and it may not be imposed upon him. The quotation does not have to be in accordance with Actual/Defined Cost plus Fee and therefore could be whatever the *Contractor* wishes to charge (always within the boundaries of mutual trust and cooperation, of course!).

If the *Project Manager* is concerned because a compensation event is pushing out the Completion Date, he may request the *Contractor* to submit alternative quotations for the compensation event in order to maintain its Completion Date, rather than instruct a quotation for acceleration. See NEC Book 3, Managing Reality: Managing the Contract, Chapter 2 on the control of time for more details.

Acceleration in the ECC does not mean speeding up the progress of the contract to achieve Completion on time. The *Project Manager* may not instruct the *Contractor* to speed up progress if he is concerned that Completion will not be achieved by the Completion Date. He may, however, instruct the *Contractor* to submit a revised programme[80] showing how he intends to make up time. Some employers do not like this departure from traditional contracts where the contractor is required to use his best endeavours and where the engineer/project manager may instruct the contractor to speed up the works. These employers add clauses allowing the project manager to make such an instruction and allowing him to disallow the costs of such progression.

(1) As a result of new instructions from a government regulatory body, the *Employer* wishes the project to finish earlier than the Completion Date stated in the contract.

(2) The *Project Manager* instructs the *Contractor* to submit a quotation to achieve Completion on 26 April 2006 instead of the Completion Date included in the contract of 10 May 2006.

(3) The *Contractor* decides that due to his resources, he will be unable to achieve Completion earlier than the Completion Date.

(4) The *Contractor* notifies the *Project Manager* within the *period for reply* that he will not be submitting a quotation for acceleration and he includes his reasons in the notification.

OR

(3) The *Contractor* reviews his costings and his programme including his resources and decides that he is able to achieve Completion before the Completion Date. Since the *Contractor* does not have to submit his quotation based on the Schedule of Cost Components, he does not have to justify the changes to the Prices. He decides to add a little extra for non-productive overtime and to cater for the delay damages that would apply if he is unable to complete on time.

[76]ECC2 clause 95.6; ECC3 clause 91.6.
[77]Clause 36.3 in Options A and B; clauses 36.3 and 36.5 (ECC3 clause 36.3) in options C and D; clauses 36.4 and 36.5(ECC3 clause 36.4) in Options E and F.
[78]Clause 36.1.
[79]Clause 36.2.
[80]Clause 32.2.

(4) The *Contractor* submits his quotation for acceleration to the *Project Manager* within the *period for reply*. His quotation comprises the changes to the Prices and the Completion Date and a revised programme. In ECC3 any changed Key Dates are also required to be shown.

(5) The *Project Manager* does not accept the quotation. No further instruction is given regarding the acceleration and there is no recourse to a compensation event for non-acceptance.[81]

OR

(5) The *Project Manager* accepts the quotation.

(6) The *Project Manager* changes the Completion Date and accepts the revised programme and in ECC3 changes Key Dates. He also changes the Prices, depending on the main option.[82]

(7) The *Project Manager* gives the instruction to accelerate.

In ECC2, if the Subcontractor in Options C, D, E and F has a proposal to accelerate, then the *Contractor* submits the proposal to the *Project Manager* for acceptance, at which point the *Project Manager* may instruct the *Contractor* to submit a proposal for acceleration.

[81] Clause 60.1(9).

[82] Only for Options A, B, C and D. For Options E and F it is the forecast of the total Actual/Defined Cost of the whole of the *works* which is changed.

Pro-forma 12: POSSESSION/ACCESS[83] CERTIFICATE

Project number:

Description:

Contract number:

Section of the *works*:

To: (The *Contractor*)		To: (The *Employer*)	
Address:		Address:	
Telephone:		Telephone:	
Fax:		Fax:	
Attention:		Attention:	

Part of the Site	**Description**	***Possession /Access Date***

Remarks

Signed for and on behalf of the *Employer*

Signature Name Date

Distribution:				

[83]ECC2 possession = ECC3 access.

1.6.15 Tests and inspections

The tests and inspections performed or watched under an ECC are only those tests or inspections that are described in the Works Information or by the applicable law.[84] The *conditions of contract* do not refer to performance tests or tests at completion or any other tests, but simply describe the procedures to be followed in carrying out the tests. It is important, therefore, that the Works Information details all the tests and inspections that the *Employer* requires to be carried out.

1.6.15.1 How the Supervisor *performs a test or inspection*

The *Supervisor* is required to notify the *Contractor* of a test or inspection before he commences the test/inspection[85] and is also required to notify the *Contractor* of the subsequent result. The *Supervisor* may also watch any test done by the *Contractor*, not only those that he is obliged to watch.[86]

The *Supervisor* is obliged to carry out the test without causing unnecessary delay to the work or to a payment that is conditional upon a test/inspection being successful.[87] The contentious word, of course, is 'unnecessary', where the *Supervisor* and the *Contractor* could have different ideas about what delay is necessary and what is not.

The *Supervisor* should therefore:

(1) Ensure that the *Employer* has provided for the test any materials, facilities and samples that he said he would provide in the Works Information.
(2) Notify the *Contractor* that he is going to perform the test, giving the *Contractor* sufficient notice to prepare.
(3) Perform the test without causing unnecessary delay to (a) the work or (b) a payment conditional upon the work.
(4) Notify the *Contractor* of the results (copying the *Project Manager* in order that he is advised of progress).
(5) If the result shows that the test has passed, the *Contractor* continues.

OR

(6) If the result shows that a Defect is present, the *Contractor* corrects the Defect.
(7) The test is repeated.
(8) The *Project Manager* assesses the cost incurred by the *Employer* in repeating the test and the *Contractor* pays the amount assessed. Contractors should note there is no requirement for the *Project Manager*'s assessment to be based on Actual/Defined Cost.

1.6.15.2 How the Project Manager *performs a test or inspection*

The *Project Manager* may not perform tests or inspections.[88] The role of checking quality, notifying Defects and performing or watching tests or inspections is the *Supervisor*'s.

1.6.15.3 How the Contractor *performs a test or inspection*

The *Contractor* is only required to notify to the *Supervisor* those tests or inspections that are required by the contract or the law. Any other tests and inspections that he chooses to do may be done according to his own procedures and need not be notified to the *Supervisor*.

This guidance is therefore only for those tests and inspections that are required by the contract or the law.

The *Contractor* is required to notify the *Supervisor* of a test or inspection before he commences the test/inspection[89] and is also required to notify the *Supervisor* of the subsequent result. The *Contractor* is required to notify the *Supervisor* in time for a test or inspection to be arranged and done before doing other work that would obstruct the test or inspection.[90] If a test or

[84] Clause 40.1.
[85] Required by the contract or the applicable law (clause 40.1).
[86] Clause 40.3.
[87] Clause 40.5.
[88] Unless he has been delegated this duty by the *Supervisor*, in which case the *Project Manager* would be performing the action as the *Supervisor* and not as the *Project Manager*.
[89] Required by the contract or the applicable law (clause 40.1).
[90] Clause 40.3.

inspection shows that work has a Defect, the *Contractor* corrects the Defect and the test or inspection is repeated.[91]

(1) If the test (or inspection) is required by the contract or the law, the *Contractor* notifies the *Supervisor* that the work is ready to be tested, making sure that the notification gives the *Supervisor* enough time to arrange to watch the test and provide any facilities or materials that the *Employer* might be required to provide for the test. The *Contractor* makes sure that he has the facilities, samples or materials that he is required to provide for the test.
(2) After doing the test at the notified time, the *Contractor* notifies the *Supervisor*, after the test, of the results.
(3) If the test showed that the *works* had a Defect, the *Contractor* corrects the Defect.
(4) If the *Project Manager* has assessed the cost to the *Employer* of repeating the test after a Defect was found, the *Contractor* pays the amount assessed, as detailed in a subsequent payment certificate.

1.6.15.4 How to search for a Defect

The *Supervisor* may instruct the *Contractor* to search if he suspects that a Defect exists,[92] whether the work is covered or not. Searching could include:

- uncovering, dismantling, recovering and re-erecting work,
- providing facilities, materials and samples for tests and inspections done by the *Supervisor* and
- doing tests and inspections, which the Works Information does not require.

It is a compensation event if a Defect is not discovered during the search.[93]

(1) The *Supervisor* instructs the *Contractor* to search and includes in the instruction where the search is to take place and what the search entails.

1.6.15.5 How to repeat a failed test

If the test or inspection shows that the work has a Defect, the *Contractor* is required to correct the Defect before the test or inspection is repeated.[94] This applies whether it is the *Supervisor* or the *Contractor* who carried out the test or inspection. The same rules apply with respect to the provision of facilities, samples and materials for testing (as stated in the Works Information) for any repeated tests as well as for the first test. In addition, the Works Information might state that the *Employer* will not provide facilities (such as water) free of charge for repeated tests.

(1) A test (or inspection) is conducted by the *Contractor*. The results shows that there is a Defect.
(2) The *Contractor* notifies the *Supervisor* of the result.
(3) The *Contractor* corrects the Defect. Note that any delays in the programme as a result of a repeat test are the responsibility of the *Contractor*.
(4) The *Contractor* arranges to carry out a repeat, notifying the *Supervisor* of the time of the test and arranging for any *Employer*-provided materials, samples or facilities.
(5) The *Contractor* carries out the repeat test.

1.6.15.6 How to deliver Plant and Materials

If the Works Information describes certain Plant and Materials that are to be tested or inspected before being delivered to the Working Areas, the *Contractor* may not bring those items to Site until the *Supervisor* has notified the *Contractor* that they have passed the test or inspection.[95]

If the *Contractor* wants to deliver Plant and Materials to the Site, he should therefore:

[91] Clause 40.4.
[92] Clause 42.1.
[93] Clause 60.1(10).
[94] Clause 40.4.
[95] Clause 41.1.

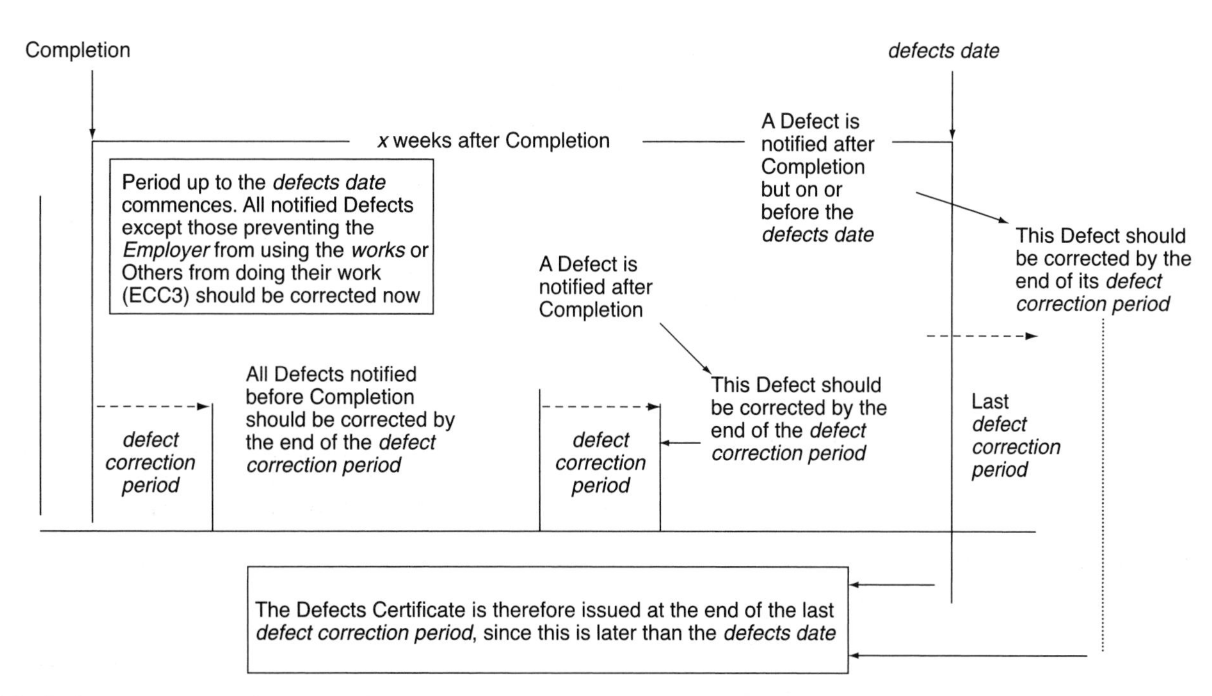

Fig. 1.3. ***Defects date*** **and** ***defect correction period***

(1) Check whether the Works Information requires them to be tested prior to coming to Site.
(2) If so, inform the *Supervisor* in advance that he is ready to deliver the items.
(3) Provide the *Supervisor* with access to the store where the Plant and Materials are held. They should be clearly marked up as being for the *works* and the Works Information will usually require proof of ownership and indemnification by the *Contractor.*
(4) Receive written notification from the *Supervisor* that the items have passed the test.
(5) Deliver the Plant and Materials to the Working Areas.

1.6.16 Defects Managing Defects is one of the *Supervisor*'s most important duties in the ECC.

A Defect has a specific meaning in the ECC[96] and does not refer to all defects or snagging. A Defect refers to the Works Information and this emphasises the importance of the Works Information to the contract.

Although the *Contractor* is required to notify Defects to the *Supervisor*, it is also the *Supervisor*'s duty to notify Defects to the *Contractor.*[97] Each Defect is required to be corrected within a certain time period,[98] the *defect correction period*, which period is identified in the Contract Data. If Defects are notified before Completion, the *Contractor* does not need to start correcting them until Completion is achieved, unless the Defect would prevent the *Employer* from using the *works* (or Others from doing their work in ECC3), in which case the Defect would have to be corrected in order to achieve Completion in the first place. Defects notified after Completion are required to be corrected within their *defect correction period*.

There may be different *defect correction period*s for different categories of Defects. In addition, some contracts may have been modified so that all Defects must be correction within their *defect correction period* after notification, whether notified before or after Completion.

[96]ECC2 clause 11.2(15); ECC3 clause 11.2(5).
[97]Clause 42.2.
[98]ECC2 clause 43.1; ECC3 clause 43.2.

The Defects Certificate[99] is issued either on the *defects date*, or at the end of the last *defect correction period*, whichever is the later.[100] The Defect Certificate signifies the end of the contract and either states that there are no Defects or contains a list of Defects that the *Contractor* has not corrected. Note that the *defects date* is calculated from Completion and not from the Completion Date.

Figure 1.3 refers.

[99]ECC2 clause 11.2(16); ECC3 clause 11.2(6).
[100]ECC2 clause 43.2; ECC3 clause 43.3.

Pro-forma 13: DEFECTS NOTIFICATION BY THE *SUPERVISOR*

Project number:

Description:

Contract number:

DEFECTS NOTIFICATION NUMBER: Date:

To: (The *Contractor*)	
Address:	
Telephone:	
Fax:	
Attention:	

The following Defects are notified:

The *defect correction period* is:			
and end of this *defect correction period* is on:			
	Day	**Month**	**Year**

Notified by *Supervisor*

Signature Name Date

Distribution:				

Pro-forma 14: DEFECTS NOTIFICATION BY THE *CONTRACTOR*

Project number:

Description:

Contract number:

DEFECTS NOTIFICATION NUMBER: Date:

To: (The *Supervisor*)	
Address:	
Telephone:	
Fax:	
Attention:	

The following Defects are notified:

The *defect correction period* is:			
and end of this *defect correction period* is on:			
	Day	**Month**	**Year**

Notified by *Contractor*

Signature Name Date

Distribution:				

Pro-forma 15: DEFECTS CERTIFICATE

Project number:

Description:

Contract number:

To: (The *Contractor*)		To: (The *Project Manager*)	
Address:		Address:	
Telephone:		Telephone:	
Fax:		Fax:	
Attention:		Attention:	

Completion achieved on:			
The *defects date* is:			
The end of the last *defect correction period* is:			
*This Defects Certificate date is:			
(*at the later of the *defects date* or the end of the last *defect correction period*)	**Day**	**Month**	**Year**

List of Defects notified before the *defects date* which the *Contractor* has not corrected (or, if there are no such Defects, a statement that there are none)	

***Works* checked by:**

Signature Name Date

Certified by *Supervisor*:

Signature Name Date

Distribution:				

Pro-forma 16: DEFECTS CORRECTED BY OTHER PEOPLE

Project number:

Description:

Contract number:

To: (The *Contractor*)		To: (The *Project Manager*)	
Address:		Address:	
Telephone:		Telephone:	
Fax:		Fax:	
Attention:		Attention:	

Refer to the *Supervisor*'s Defects Notification Dated: **Number:**

Defect reference	**End of the *defect correction period***	**Uncorrected defects**	**Assessed cost of having the Defect corrected by other people (excl. VAT)**

Checked by *Supervisor*:

Signature Name Date

Certified by *Project Manager*:

Signature Name Date

Distribution:				

1.6.16.1 How to notify a Defect

Defects are defined in the contract as a part of the *works* that is not in accordance with the Works Information.[101] In other words, if the requirement is not included in the Works Information as something that the *Contractor* is to provide, not providing it cannot be a Defect. Every defect need not be a Defect and instructing the *Contractor* to correct something that is not a Defect (although it could be a defect) is in fact a compensation event.

1.6.16.2 How the Contractor *notifies a Defect*

(1) The *Contractor* notices that there is a Defect in work that has just been carried out.
(2) He completes a Defect notification form and sends it to the *Supervisor*.
(3) If the Defect is one that would prevent the *Employer* from using the *works* (or Others from doing their work, in ECC3), the *Contractor* corrects the Defect prior to Completion (in practical terms, this would probably mean immediately).
(4) If the Defect is not one that would prevent the *Employer* from using the *works*, the *Contractor* can choose whether to correct the Defect immediately (this would normally be the most efficient and effective time to do it), or whether to wait until after Completion. If the *Contractor* waits until after Completion, he has only the time of the *defect correction period* to correct the Defect and any others which he is required to complete after Completion.

1.6.16.3 How the Supervisor *notifies a Defect*

(1) The *Supervisor* notices that there is a Defect in work that has just been carried out.
(2) He completes a Defect notification form and sends it to the *Contractor*. On the Defect notification, the *Supervisor* would state whether he considers that the Defect is one that would prevent the *Employer* from using the *works* (or Others from doing their work, in ECC3) and therefore when the Defect should be corrected by.
(3) The *Supervisor* would file the Defect notification with the others and use them when issuing the Defects Certificate.

1.6.16.4 How the Project Manager *notifies a Defect*

It is the duty of the *Supervisor* to notify a Defect. Any communication from the *Project Manager* notifying a Defect has no efficacy under the contract.

1.6.16.5 How to correct a Defect

The *defect correction period* is not to be confused with the *defects date* or the defects liability period or maintenance period in traditional contracts.

The *defect correction period* is the maximum period within which a *Contractor* must correct a notified Defect. It is not the whole period from Completion to the *defects date* but a period that the *Employer* may have inserted in Contract Data part one, say two weeks. Each Defect must be corrected within this period, from its notification, or after Completion.

If a Defect is notified before Completion and it does not prevent and *Employer* from using the *works*[102] (or Others from doing their work, in ECC3), then the Defect must be corrected at Completion[103] and it must be corrected within its *defect correction period*, say two weeks. Presumably if there are many Defects, the *Contractor* will have to start the correction process before Completion if he is to correct them all within the two weeks.

Defects that are notified after Completion must still be corrected within their *defect correction period*, say two weeks, but the period starts when the Defect is notified.[104]

There are two further aspects to the *defect correction period* that require some discussion.

[101] ECC2 clause 11.2(15); ECC3 clause 11.2(5).
[102] If the Defect prevents the *Employer* from using the *works*, then Completion cannot be achieved (clause 11.2(13); ECC3 clause 11.2(2)).
[103] ECC2 clause 43.1; ECC3 clause 43.2.
[104] ECC2 clause 43.1; EEC3 clause 43.2.

First, employers of maintenance contracts may wish to amend the core clause that requires Defects to be corrected only after Completion. Many contracts require Defects to be corrected at the time of notification, for example leaving out a danger sign for an overhead line should be immediately corrected. In this case, clause 43.1[105] would be amended via an Option Z clause to read 'This period begins when the Defect is notified for all Defects.' If the contract is an Option C, D or E contract, the Disallowed Cost clause should also be amended via an Option Z clause so that the bullet point reading 'correcting Defects after Completion' instead reads 'correcting Defects'.

Second, different *defect correction periods* may be introduced for different categories of Defects. ECC3 faciliates this in Contract Data part one. ECC2 requires categories to be stated in the Contract Data and the categories described in the Works Information. This may be useful where some Defects require rectification immediately because they would inhibit the functioning of the *works* while other, more cosmetic Defects may be corrected within two or four weeks. Of course, the key is to describe them carefully in the Works Information and to state in the Contract Data that this is where the descriptions are to be found.

In general, however, careful thought should be invested in the *defect correction period*. Remember also that anything can be changed by negotiation and, if the *Contractor* finds that a particular *defect correction period* is too tight regarding a large Defect, the *Project Manager* would be advised to allow the *Contractor* more time to correct the Defect properly.

Various incentive schemes have been used regarding Defects. Although the *Contractor* is supposed to notify his own Defects, traditionally contractors have not done so and may have been happy to get away with not correcting Defects that have not been spotted by the employer and his engineer. In order to encourage the *Contractor* to notify and then correct his own Defects, some contracts disallow the costs of Defects notified by the *Supervisor* but allow the costs of Defects notified by the *Contractor.*[106] Others split the cost of correcting Defects between the two Parties equally. In Options C, D and E, this may simply encourage the *Contractor* not to notify Defects but correct them anyway, since the cost of such correction would be paid as an Actual Cost/Defined Cost unless noticed by the *Project Manager* or the *Supervisor*.

(1) A Defect has been notified (whether by the *Contractor* or the *Supervisor*). It is not a Defect that would prevent the *Employer* from using the *works* (or Others from doing their work in ECC3).
(2) The *Contractor* decides that it would be more efficient to correct the Defect concurrently with the work he is currently doing and he has the float within his programme to enable him to do this. He corrects the Defect immediately.
(3) Since he has until the *defect correction period* after Completion to correct the Defect, his obligation is complete.

OR

(1) A Defect has been notified (whether by the *Contractor* or the *Supervisor*). It is not a Defect that would prevent the *Employer* from using the *works* (or Others from doing their work in ECC3).
(2) The *Contractor* decides that the Defect is so minor as to be cosmetic and that he does not have the luxury within his programme to spend the time correcting the Defect before Completion.
(3) The *Contractor* must correct the Defect within the *defect correction period* after Completion.

OR

(1) A Defect has been notified (whether by the *Contractor* or the *Supervisor*). It is a Defect that would prevent the *Employer* from using the *works* (or Others from doing their work in ECC3).

[105] ECC3 clause 43.2.
[106] This will not work with Option A and B contracts.

(2) The *Contractor* is required to correct the Defect before Completion and he decides to correct the Defect immediately.

OR

(1) A Defect has been notified (whether by the *Contractor* or the *Supervisor*). It is a Defect that would prevent the *Employer* from using the *works* (or Others from doing their work in ECC3).
(2) The *Contractor* is required to correct the Defect before Completion.
(3) He does not correct the Defect.
(4) The *Project Manager* assesses the cost of having the Defect corrected by other people and the *Contractor* pays that amount.

OR

(1) A Defect has been notified after Completion.
(2) The *Contractor* is required to correct the Defect within the *defect correction period*.
(3) The *Employer* allows the *Contractor* access to the Defect.
(4) He does not correct the Defect.
(5) The *Project Manager* assesses the cost of having the Defect corrected by other people and the *Contractor* pays that amount.[107]

OR

(1) A Defect has been notified after Completion.
(2) The *Contractor* is required to correct the Defect within the *defect correction period*.
(3) The *Employer* does not allow the *Contractor* access to the Defect due to operational reasons.
(4) The *Contractor* cannot correct the Defect.
(5) The *Project Manager* assesses how much it would have cost the *Contractor* to correct the Defect[108] and the *Contractor* pays that amount.[109]

1.6.16.6 How to issue the Defects Certificate

There is no defects liability period or maintenance period in the ECC, *per se*. The period within which the *Contractor* is obliged to correct Defects free of charge is the period between Completion of the whole of the *works* and the *defects date*. The *defects date* is stated in the Contract Data as being a number of weeks after Completion of the whole of the *works*, usually 52 weeks, sometimes 26 weeks, on process plant contracts it may be 104 weeks (two years), it depends entirely on the *works*. Since the *defects date* is dependent on Completion, it does not matter whether Completion is before, on or after the Completion Date, since the *defects date* runs from Completion and not the Completion Date.

The *defects date* serves three purposes in the ECC:

(1) It is the last date by which either the *Supervisor* or the *Contractor* can notify Defects.[110]
(2) It is the date on which the *Supervisor* issues the Defects Certificate (unless a Defect notified before the *defects date* has a *defect correction period* that ends later than the *defects date*, in which case the Defects Certificate is issued on the later date).[111]
(3) It sets the final date for the notification of compensation events.[112]

The *Supervisor* is required to issue the Defects Certificate at the later of the *defects date* and the last *defect correction period*. This means that if a Defect was notified two days before the *defects date* and the *defect correction period*

[107] ECC3 clause 45.1.
[108] Note that this is different from assessing how much it would cost someone else to correct the Defect.
[109] ECC3 clause 45.2.
[110] Clause 42.2.
[111] ECC2 clause 43.2; ECC3 clause 43.3.
[112] Clause 61.7.

within which that Defect must be corrected is two weeks, then the *Supervisor* can only issue the Defects Certificate 12 days after the *defects date*.

(1) The *Supervisor* notifies a Defect to the *Contractor* seven days before the *defects date*. The *defect correction period* is 14 days.
(2) The *Contractor* corrects the Defect within eight days (that is, one day after the *defects date*).
(3) The *Supervisor* issues the Defects Certificate seven days after the *defects date* with a statement that there are no Defects.

OR

(1) The *Supervisor* notifies a Defect to the *Contractor* seven days before the *defects date*. The *defect correction period* is 14 days.
(2) The *Contractor* corrects the Defect within nine days after the *defects date* (that is, two days longer than the *defect correction period*).
(3) The *Supervisor* issues the Defects Certificate seven days after the *defects date* with a statement that there are still Defects and a list of the existing Defects, including the one that was not corrected within 14 days.

OR

(1) No Defects are notified within a six-month period prior to the *defects date*.
(2) The *Contractor* has corrected all previously notified Defects.
(3) The *Supervisor* issues the Defects Certificate on the *defects date* with a statement that there are no Defects.

OR

(1) No Defects are notified within a six-month period prior to the *defects date*.
(2) The *Contractor* has not been on Site since one month after Completion and has not corrected any Defects notified after Completion.
(3) The *Supervisor* issues the Defects Certificate on the *defects date* with a statement that there are Defects and a list of those Defects.
(4) The *Project Manager* assesses the cost of having the Defect corrected by other people and the *Contractor* pays that amount. The *Employer* may use any retention to employ others to correct the Defects, or to correct them himself.

1.6.16.7 How to accept a Defect

Either the *Contractor* or the *Project Manager* may propose to the other to accept a Defect.[113] Note that it is still the *Supervisor* who notifies the Defect, but the *Project Manager* may choose to accept a Defect where he considers that it is more effective to do so. If both the *Contractor* and the *Project Manager* agree, the *Contractor* submits a quotation of reduced Prices, an earlier Completion Date or both to the *Project Manager* for acceptance. If the *Project Manager* decides to accept the quotation, then he gives an instruction to change the Works Information accordingly.[114]

1.6.16.8 How the Contractor *proposes to accept a Defect*

(1) The *Supervisor* notifies a Defect to the *Contractor*.
(2) The *Contractor* suspects that the Defect will not affect the *works* unduly and so he proposes to the *Project Manager* that the Works Information should be changed so that the Defect does not have to be corrected.
(3) If the *Project Manager* is prepared to accept a change, he instructs the *Contractor* to submit a quotation for reduced Prices, an earlier Completion Date or both.
(4) The *Contractor* submits the quotation within the *period for reply*.
(5) The *Project Manager* gives his decision to the *Contractor* within the *period of reply* of receiving the quotation.

[113]Clause 44.1.
[114]Clause 44.2.

(6) If his decision is not to accept the Defect, the *Contractor* must correct the Defect within the *defect correction period* after the notification of the decision.
(7) If his decision is to accept the Defect, the *Project Manager* gives an instruction to change the Works Information, the Prices and the Completion Date.
(8) The instruction is not a compensation event.[115]

1.6.16.9 How the Project Manager *proposes to accept a Defect*

(1) The *Contractor* notifies a Defect to the *Supervisor* who copies in the *Project Manager*.
(2) The *Project Manager* decides that the Defect will not inconvenience the *Employer* disproportionately so he proposes to the *Contractor* that the Works Information should be changed so that the Defect does not have to be corrected.
(3) If the *Contractor* is prepared to accept a change, he informs the *Project Manager.*
(4) The *Project Manager* instructs the *Contractor* to submit a quotation for reduced Prices, an earlier Completion Date or both.
(5) The *Contractor* submits the quotation within the *period for reply.*
(6) The *Project Manager* gives his decision to the *Contractor* within the *period of reply* of receiving the quotation.
(7) If his decision is not to accept the Defect, the *Contractor* must correct the Defect within the *defect correction period* after the notification of the decision.
(8) If his decision is to accept the Defect, the *Project Manager* gives an instruction to change the Works Information, the Prices and the Completion Date.
(9) The instruction is not a compensation event.[116]

1.6.17 How to repair the *works*

Until the Defects Certificate has been issued, the *Contractor* is under an express obligation to replace promptly losses or to repair damage to the *works*, Plant and Materials.[117]

This obligation must be fulfilled whether or not the *Project Manager* has given an instruction to that effect and it relates not only to repairs arising from the *Contractor*'s own actions or risks but also those of the *Employer*'s actions which are at his risk under clause 80.1.

Essentially the *Contractor*'s obligation is to construct the *works.* Should the *works* be damaged, destroyed, etc., the *Contractor* is not relieved from the performance of the contract unless such losses arise from a fault of the *Contractor* in which case the *Employer* may terminate under clause 95.2 (ECC3 clause 91.2). This clause has been included so that the contract would not become frustrated in legal terms.

If the *Contractor* fails to carry out his obligation to repair, then the *Employer* will be able to claim for damages for breach of contract (not belonging to the Contractor).

(1) A large crane rumbles by a structure being built by the *Contractor*. The crane causes excessive vibration to the structure and accidentally cracks the windows of the structure.
(2) The *Project Manager* does not instruct the *Contractor* not to repair the windows.
(3) The *Contractor* promptly repairs the damage.
(4) The *Contractor* notifies a compensation event under clause 60.1(14).
(5) The compensation event procedure continues.

The ECC is not clear on how and on what basis the *Contractor* is paid for repair *works*, which are not his fault or responsibility. On the basis that it is not a *Contractor*'s risk then it must be an *Employer*'s risk under clause 80.1 and therefore a compensation event under clause 60.1(14) *Employer*'s risks.

[115] Clause 60.1(1).
[116] Clause 60.1(1).
[117] Clause 82.1.

1.6.18 Payment The *Project Manager* is required to assess the amount due, with or without an application for payment by the *Contractor*. The *Project Manager* includes any monies due to or from the *Contractor* and notifies the *Contractor* through certification. The *Employer* pays the *Contractor* within a period of time after the assessment date **or** within a period of time after the certification (if Option Y(UK)2 has been chosen).

1.6.18.1 How to make an application for payment

(1) The *Contractor* does not assess the amount due since this is an obligation of the *Project Manager*. The *Contractor* may, however, submit an application for payment before the assessment date. The *Contractor* applies for what he considers to be the amount due to him either on or before the assessment date. It would assist the *Project Manager* more if it were submitted before the assessment date.

(2) In his assessment, the *Contractor* takes into account all the things that the *Project Manager* would have taken into account.

1.6.18.2 How the Project Manager *assesses the amount due*

(1) The *Project Manager* receives an application for payment five days before the assessment date.

(2) The *Project Manager* calculates the Price for Work Done to Date. This would depend on the main Option applicable to the contract.

(3) The *Project Manager* also takes into account the other secondary Options applicable to the contract if they affect payment, for example retention, delay damages, advanced payments, bonus for early Completion.

(4) The *Project Manager* takes into account any other amounts due to or from the *Contractor*.

(5) If the *Contractor* has not submitted a first programme for acceptance showing the information required by the contract, the *Project Manager* withholds 25% of the Price for Work Done to Date.

(6) The *Project Manager* adds any interest due to the *Contractor* for late payments, late certificates or incorrect amounts due.

1.6.18.3 How to certify an amount due

(1) Having assessed the amount due, the *Project Manager* includes the final amount and its total breakdown in a pro-forma certificate and issues it to the *Contractor* within seven days of the assessment date.

(2) For Options C, D and E contracts where the assessment of the amount due is more challenging, the *Project Manager* could do a spot check and reserve a comprehensive check for audit.

1.6.18.4 How to pay an amount due

(1) The *Employer* is obliged to make payment within three weeks of the assessment date (or another period stated in the Contract Data). Any late payments are subject to the payment of interest to the *Contractor* as assessed in the next amount due.

(2) If the contract is subject to the Housing Grants, Construction and Regeneration Act 1996, the *Employer* is obliged to pay within three weeks after the date on which payment becomes due (or another period stated in the Contract Data). The date on which payment becomes due is seven days after the assessment date and would usually coincide with the payment certificate issued by the *Project Manager*. Any late payments are subject to the payment of interest to the *Contractor* as assessed in the next amount due.

1.6.18.5 How to withhold payment

(1) The *Employer* decides that he intends to withhold payment of an amount due under the contract. A Party does not withhold payment of an amount due under this contract unless he has notified his intention to do so.

(2) Not later than seven days before the final date for payment, the *Employer* notifies the *Contractor* of this intention, specifying the amount to be withheld and the grounds for withholding it.

Pro-forma 17: PAYMENT CERTIFICATE

Project number:

Description:

Contract number:

To: (The *Contractor*)		To: (The *Project Manager*)	
Address:		Address:	
Telephone:		Telephone:	
Fax:		Fax:	
Attention:		Attention:	

Certificate No:	
Assessment date:	
Certificate date:	
Certificate due date:	
Payment due date:	
Prepared by:	

Total amount due to/from the *Contractor*	

Certified by *Project Manager*:

Signature Name Date

Distribution:				

Pro-forma 17: PAYMENT CERTIFICATE (Cont'd)

(a)	Cumulative Price for Work Done to Date (excl. VAT)				
(b)	Retention at 25% of (a) if no Accepted Programme				
(c)	**Subtotal (a) – (b)**				
(d)	Option P/Option X16 – Retention				
	(i) retention free amount				
	(ii) retention percentage				
	Amount retained [(a – d(i)) × d(ii)]				
(e)	**Subtotal (c) – (d)**				
(f)	Retention released at Completion/Defects Certificate				
(g)	**Subtotal (e) + (f)**				
(h)	Interest:				
	(i) late payment				
	(ii) corrected amounts – underpayment				
	(iii) corrected amounts – overpayment				
	(iv) late certification (ECC2 only)				
	Total interest [h (i) + (ii) + (iv) – (iii)]				
(j)	Option Q/Option X6 – bonus for early Completion @ £ ... /day for ... days				
(k)	Option R/Option X7 – damages @ £ ... /day for ... days				
(l)	Option S/Option X17 – low-performance damages @ [formula]				
(m)	Total bonus/damages due to *Contractor* [j – k + l]				
	Total bonus/damages due to *Employer* [j – k + l]				
(n)	**Total cumulative amount certified [g + h + m]**				
(p)	**Less amount previously certified**				
(r)	**Amount due on this certificate**				
(s)	**Add VAT @ 17.5%**				
(t)	**Total amount due to/from *Contractor***				

Assessment checked by:

Signature   Name   Date

Certified by *Project Manager*:

Signature   Name   Date

Distribution:				

1.6.19 Compensation events

Compensation events are those events for which the *Contractor* becomes entitled to an assessment of time and money. Compensation events tend to be a contractual remedy to the *Project Manager*'s or the *Employer*'s breach of contract.

There is no such thing as a 'variation', an 'extension of time', or 'delay and disruption' under the ECC. There is also no concept of a 'claim'. The new terminology of a compensation event will hopefully reduce the instance of buying contracts, change the claims mentality existing in some parts of the industry and make clients more aware that changes cost money.

The compensation event procedure is designed to deal with changes and other factors as they occur, to encourage good management and decrease conflict. It is also designed to keep the *Contractor* risk-neutral in a compensation event situation.

Either the *Project Manager* or the *Contractor* may notify a compensation event.

The *Project Manager* would usually notify a compensation event where the compensation event has resulted from either himself or the *Supervisor* instructing something or changing a decision. For example, at the same time as he instructs a change to the Works Information, the *Project Manager* should notify a compensation event. If, however, the *Contractor* considers that the *Project Manager* or the *Supervisor* have given an instruction but the *Project Manager* has not notified a compensation event, then the *Contractor* should, in his own interests, notify the compensation event.

In ECC2 the *Contractor* has only two weeks to notify a compensation event after becoming aware of it. It was always the intent of the drafters that the two-week period would be a time-bar to any time or money entitlement. This has been the subject of much debate and discussion and the period has now been extended to eight weeks in ECC3 on the basis that if it is to be a time-bar then it should be accompanied by a reasonable time-scale. The trade-off of having an extended period is that there may be a loss of certainty of outcome. It will be for *Employers* and *Contractors* to consider the implications of this change and what period of time is considered as being reasonable. Once he has notified a compensation event, the *Contractor* has to wait for the *Project Manager* to assess the validity of the compensation event before receiving either notification that the Prices and the Completion Date (and Key Date in ECC3) will not be changed (where the *Project Manager* considers that the compensation event is not valid) or an instruction to submit quotations for the event.

The pro-formas (which follow section 1.6.19.8 below) include the following:

- compensation event notification by the *Project Manager*,
- compensation event notification by the *Contractor*,
- *Project Manager*'s reply to a compensation event notification by the *Contractor*,
- *Project Manager*'s reply to a compensation event quotation by the *Contractor*,
- instruction to submit quotations for a proposed instruction or proposed changed decision.

The last pro-forma listed above refers to the possibility that the *Project Manager*, on considering a decision that could lead to a compensation event, first wants to ascertain from the *Contractor* what the cost of that change or decision will be. He therefore instructs the *Contractor* to submit quotations for a proposed change[118] and then either notifies a compensation event if the quotation is acceptable, or does not notify a compensation event.

Although the first two pro-formas refer to the 18 (19 in ECC3) compensation events in clause 60.1, the form may require some adjusting in order to cater for the additional compensation events in some main and secondary Options and the Contract Data.

[118] Clause 61.2.

1.6.19.1 How the Contractor *notifies a compensation event*

(1) If the *Contractor* answers yes to **all** of the following, then he may notify a compensation event to the *Project Manager*:
- An event has happened or you expect it to happen.
- You believe the event is a compensation event.
- It is less than two weeks (eight weeks in ECC3[119]) since you became aware of the event.
- The *Project Manager* has not notified the event to you.

(2) The *Contractor* includes in his notification a description of the event and the clause number relating to the compensation event he believes it is.

(3a) ECC2 only. The *Contractor* waits one week (or a longer period to which the *Contractor* has agreed) for the *Project Manager*'s decision regarding the notification or an instruction to submit quotations.

(3b) ECC3 only. If the *Project Manager* does not notify his decision to the *Contractor* within one week or a longer period to which the *Contractor* has agreed, then the *Contractor* may notify the *Project Manager* to this effect. If the *Project Manager* fails to reply within two weeks of this notification, the non-response is treated as acceptance by the *Project Manager* that the event is a compensation event and an instruction to submit quotations. In this case, the *Contractor* is advised to include this clause on his quotation when he submits quotations.

This change has been introduced to cater for the non-performance of *Project Managers* on contracts. Certainly many *Contractors* have experienced problems with *Project Managers* who simply do not respond within the time-scales. A side-effect may be that *Employers* may be encouraged to delete or extend the time period. It may also encourage greater use of requesting a longer period to reply, but this is subject to the *Contractor*'s agreement.

1.6.19.2 How the Project Manager *notifies a compensation event*

(1) If the *Project Manager* or the *Supervisor* do any of the following, then the *Project Manager* is obliged to notify a compensation event:

- The *Project Manager* gives an instruction changing the Works Information.
- The *Project Manager* gives an instruction to stop or not to start any works.
- The *Project Manager* gives an instruction for dealing with an object of value or of historical or other interest found within the Site.
- The *Project Manager* or the *Supervisor* changes a decision previously communicated to the *Contractor.*
- The *Supervisor* instructs the *Contractor* to search and no Defect is found.
- The *Project Manager* certifies take over before both Completion and the Completion Date.
- The *Project Manager* notifies a correction to an assumption about the nature of a compensation event.

(2) The *Project Manager* notifies the compensation event at the time of the event. For those events where he only finds out afterwards (such as no Defect being found after a search), he notifies a compensation event as soon as he finds out.

(3) He also instructs the *Contractor* to submit quotations for the event.

(4) He notifies the compensation event by issuing a compensation event notification to the *Contractor*.

1.6.19.3 How to submit a quotation for a compensation event

(1) The *Project Manager* instructs the *Contractor* to submit quotations (either at the same time as the notification of the compensation event or within a week after receiving the *Contractor*'s notification of a compensation event).

(2) The *Contractor* has three weeks to submit his quotation.

[119]Note that if more than eight weeks have passed and it is a compensation event which the *Project Manager* should have notified, the *Contractor* does not lose his entitlement to a compensation event.

(3) His quotation comprises changes to the Prices and the programme as affected by the compensation event. In ECC3 it also includes Key Dates.

(4) The quotation should contain enough detail so that the *Project Manager* can assess it without recourse to the *Contractor* for more information.

1.6.19.4 How the Contractor *assesses a compensation event*

(1) The *Contractor*'s quotation comprises changes to the Prices and to the programme and in ECC3 changes to Key Dates.

(2) A revised programme (called alterations to the Accepted Programme in ECC3) is submitted as part of the quotation if any part of the programme has been affected by the compensation event, for example the planned Completion Date, the method statement,[120] the resource statement, or the risk allowances.

(3) The changes to the Prices comprise the effect of the compensation event upon the Actual Cost/Defined Cost of the work already done, the forecast Actual Cost/Defined Cost of the work not yet done and the resulting Fee.

(4) The changes to the Prices are the changes to the *activity schedule* (Option A or C) or the *bill of quantities* (Option B or D). The changes are reflected in the lump sum prices or rates attached to the *activity schedule* (Option A or C) or the *bill of quantities* (Option B or D) respectively. For Options C and D, the changes to the Prices amend the target cost, thereby bringing the target cost in line with the Price for work done to Date.

1.6.19.5 How the Project Manager *assesses a compensation event*

(1) The *Project Manager* assesses the *Contractor*'s quotation by using the tools that are available to him, such as the Accepted Programme, as well as his own experience.

1.6.19.6 How to accept a quotation for a compensation event

(1) If he decides to accept the quotation, the *Project Manager* notifies the *Contractor* of his acceptance within two weeks of receiving the *Contractor*'s quotation.

(2) In ECC3, if the *Project Manager* does not reply to a quotation from the *Contractor*, then this non-response is deemed to be acceptance of the quotation.[121]

1.6.19.7 How the Project Manager *implements a compensation event – ECC2 only*

(1) If the *Project Manager* has accepted the *Contractor*'s quotation and the compensation event has already occurred, he implements the compensation event by notifying the *Contractor* of his acceptance of the quotation.

(2) If the *Project Manager* has accepted the *Contractor*'s quotation and the compensation event has not yet occurred (for example, when he instructed a quotation for a proposed instruction), he implements the compensation event when the compensation event occurs.

(3) If the *Project Manager* decides to make his own assessment and the compensation event has already occurred, he implements the compensation event by notifying the *Contractor* of his own assessment.

(4) If the *Project Manager* decides to make his own assessment and the compensation event has not yet occurred (for example, when he instructed a quotation for a proposed instruction), he implements the compensation event when the compensation event occurs.

1.6.19.8 How the Project Manager *implements a compensation event – ECC3 only*

(1) If the *Project Manager* has accepted the *Contractor*'s quotation, the compensation event is implemented when the *Project Manager* notifies the *Contractor* that he has accepted the quotation.

(2) If the *Project Manager* decides to make his own assessment, the compensation event is implemented when the *Project Manager* notifies the *Contractor* of his assessment.

(3) If the *Project Manager* did not respond to a *Contractor*'s quotation for a compensation event, then the *Contractor* may notify the *Project Manager* that he (the *Project Manager*) has not replied within the required time and which quotation is deemed to have been accepted. The compensation event is implemented when the *Contractor*'s notification is received.

[120] Called a 'statement of how the *Contractor* plans to do the work' in ECC3.
[121] ECC3 clause 62.6.

Pro-forma 18: COMPENSATION EVENT NOTIFICATION BY THE *PROJECT MANAGER*

Project number:

Description:

Contract number:

Compensation event number:

To: (The *Contractor*)	
Address:	
Telephone:	
Fax:	
Attention:	

The compensation event is as per clause 60.1 sub-paragraph:

1	2	3	4	5	6	7	8	9	10	11	12	13	14	15	16	17	18	19

	Day	Month	Year
The event started to occur on:			

The details of the event are as follows:

	Day	Month	Year
The date of this notification by the *Project Manager* is (at the same time as the occurrence of the event):			

	Yes	No
Quotations for the event have already been submitted		
The *Project Manager* hereby instructs the *Contractor* to submit quotations		
The *Project* Manager hereby notifies the *Contractor* that he did not give an early warning for this event, which he could have given		

The *Project Manager* hereby states the following assumptions about the event, because the effects of the event are too uncertain to be forecast reasonably . . . .

The *Project Manager* hereby instructs the *Contractor* to submit alternative quotations based on the following ways of dealing with the event . . . .

	Day	Month	Year
The date by which the *Contractor* is to submit quotations is (within three weeks of being instructed to do so)			

Notified by *Project Manager*:

Signature

Name

Date

Distribution:				

Pro-forma 19: COMPENSATION EVENT NOTIFICATION BY THE *CONTRACTOR*

Project number:

Description:

Contract number:

Compensation event number:

To: (The *Project Manager*)	
Address:	
Telephone:	
Fax:	
Attention:	

The compensation event is as per clause 60.1 sub-paragraph:																		
1	**2**	**3**	**4**	**5**	**6**	**7**	**8**	**9**	**10**	**11**	**12**	**13**	**14**	**15**	**16**	**17**	**18**	**19**

	Day	Month	Year
The event started to occur on:			

The details of the event are as follows:

	Day	Month	Year
The date of this notification is (not more than two weeks after the *Contractor* became aware of it, in ECC3 eight weeks)			

Notified by *Contractor*:

Signature Name Date

Distribution:				

Pro-forma 20: ***PROJECT MANAGER*'S REPLY TO A COMPENSATION EVENT NOTIFICATION BY THE *CONTRACTOR***

Project number:

Description:

Contract number:

Compensation event number: Date:

To: (The *Contractor*)	
Address:	
Telephone:	
Fax:	
Attention:	

The *Project Manager* decides that the event notified by the *Contractor*	Yes	No
• arose from a fault of the *Contractor*		
• has not happened and is not expected to happen		
• has no effect on Actual Cost/Defined Cost or Completion		
• is not one of the compensation events stated in this contract		
• was notified more than two weeks EEC2/eight weeks EEC3 after the *Contractor* became aware of it		

	Yes	No
The *Project Manager* hereby notifies the *Contractor* that the Prices and the Completion Date will not be changed – ECC3 Key Date		
The *Project Manager* hereby instructs the *Contractor* to submit quotations for the event		
The *Project* Manager hereby notifies the *Contractor* that he did not give an early warning for this event, which he could have given		
The *Project Manager* hereby states the following assumptions about the event, because the effects of the event are too uncertain to be forecast reasonably		
The *Project Manager* hereby instructs the *Contractor* to submit alternative quotations based on the following ways of dealing with the event		

The date by which the *Contractor* is to submit quotations is (within three weeks of being instructed to do so)	Day	Month	Year

Notified by *Project Manager*:

Signature Name Date

Distribution:				

Pro-forma 21: ***PROJECT MANAGER*'S REPLY TO A QUOTATION FOR A COMPENSATION EVENT**

Project number:

Description:

Contract number:

Compensation event number: Date:

To: (The *Contractor*)	
Address:	
Telephone:	
Fax:	
Attention:	

			Day	Month	Year
The *Contractor*'s quotation was received on					
The date of this reply is (within two weeks of the submission of the quotation unless extended by agreement)					
The *Project Manager*'s reply to the quotation is:				Yes	No
• acceptance of the quotation					
• a notification that the proposed instruction or a proposed changed decision will not be given					
• a notification that the *Project Manager* will be making his own assessment for the following reasons	Yes	No			
• the *Contractor* did not submit the required quotation and details of his assessment within the time allowed					
• the *Contractor* has not assessed the compensation event correctly					
• the *Contractor* has not submitted a programme which this contract requires him to submit					
• the *Contractor*'s latest programme has not been accepted for one of the reasons stated in this contract					
• an instruction to submit a revised quotation					
The reasons for instructing the *Contractor* to submit a revised quotation are as follows:					

Notified by *Project Manager*:

Signature Name Date

Distribution:				

Pro-forma 22: INSTRUCTION TO SUBMIT QUOTATIONS FOR A PROPOSED INSTRUCTION OR PROPOSED CHANGED DECISION

Project number:

Description:

Contract number:

Number: Date:

To: (The *Contractor*)	
Address:	
Telephone:	
Fax:	
Attention:	

In terms of clause 61.2, the *Project Manager* hereby instructs the *Contractor* to submit quotations for the following proposed instruction or proposed changed decision:			
The *Project Manager* hereby states the following assumptions about the event, because the effects of the event are too uncertain to be forecast reasonably			
The *Project Manager* hereby instructs the *Contractor* to submit alternative quotations based on the following ways of dealing with the event			
Quotations comprise proposed changes to the Prices and any delay to the Completion Date, ECC3 Key Dates			
The *Contractor* submits details of his assessment with his quotation.			
If the programme for the remaining work is affected, the *Contractor* includes a revised programme with his quotation showing the effect.			
	Day	Month	Year
The date of this notification by the *Project Manager* is			
The date the *Contractor* is to submit quotations is (within three weeks of being instructed to do so)			

Notified by *Project Manager*:

Signature Name Date

Distribution:				

1.6.20 Title

1.6.20.1 How to mark goods for title

If the *Supervisor* has marked goods for the contract, the title of those goods passes to the *Employer* even if they are outside the Working Areas.[122] The *Supervisor* would mark the goods if the contract identifies them for payment and if the *Contractor* has prepared them for marking as required by the Works Information.[123]

Title of goods and Equipment[124] passes in any case to the *Employer* as soon as they are brought within the Working Areas.[125]

The steps involved in marking goods for title are therefore as follows:

(1) Are the Plant and Materials outside the Working Areas?
(2) Does the contract identify Plant and Materials for payment?
(3) Has the *Contractor* prepared the Plant and Materials for marking as required by the Works Information?
(4) The *Supervisor* marks the Plant and Materials for the contract.
(5) Title passes to the *Employer*.

1.6.21 Risk

1.6.21.1 How the Contractor *assesses his risk under the contract*

The *Contractor* has both financial and project risk for any project.

Financial risk

The *Contractor*'s financial risk under the contract is determined by the main and possibly by the secondary Options chosen for the contract. For example, the *Contractor*'s financial risk is greater for Option A than for Option B because Option A is a lump sum contract and Option B is a remeasurable contract.

Project risk

The *Contractor*'s project risk is determined by a number of things, such as:

- How many changes there might be to the Works Information.
- How reliable is the Site Information?
- What are the delay damages on the contract?
- How reliable is the labour and Equipment used?
- Quality of site team.
- What is the availability of labour?
- What is the availability of Equipment?
- What is the availability of Plant and Materials?
- Might there be adverse weather that lies within the *Contractor*'s risk?
- How much float was the *Contractor* able to build into his programme?
- How fair and trustworthy are the *Project Manager* and the *Supervisor*?
- How many projects of this sort has the *Contractor* done before?
- What external influences are there, for example shareholders?
- Who are the project stakeholders both internal and external?

1.6.21.2 How the Employer *assesses his risk under the contract*

The *Employer* also experiences financial and project risk under the contract.

Financial risk

Under an Option E, the *Employer* adopts much more financial risk than under Option A. If the *Employer* can tighten up his Works Information, it is to his advantage to try to contract under Option A.

Project risk

The project risk assessed by the *Employer* will follow similar lines as the *Contractor*.

1.6.22 Insurance

Both the *Employer* and the *Contractor* are required to give to the other Party the policies and certificates for the insurance to be provided.[126] The other Party may insure a risk if the Party who is supposed to insure has not presented certificates for acceptance.

[122]Clause 70.1.
[123]Clause 71.1.
[124]Not Equipment in ECC3, only Plant and Materials.
[125]Clause 70.2.
[126]Clause 85.1 and 87.1.

1.6.22.1 How the Employer *submits policies and insurance*

(1) The *Employer* includes in the Contract Data any insurance that he intends to provide.
(2) The *Project Manager* submits the policies and certificates for insurances provided by the *Employer* to the *Contractor* for acceptance before the *starting date* and afterwards as the *Contractor* instructs.
(3) The *Contractor* accepts the policies and certificates if they comply with the contract.
(4) If the *Employer* does not submit a required policy or certificate, the *Contractor* may insure a risk that the contract requires the *Employer* to insure.
(5) The *Employer* pays the cost of this insurance to the *Contractor.*
(6) The Parties comply with the terms and conditions of the insurance policies.

1.6.22.2 How the Contractor *submits policies and insurance*

(1) The *Employer* includes in the Contract Data the insurance that he requires the *Contractor* to provide.
(2) The *Contractor* submits the policies and certificates for insurances that he provides to the *Project Manager* for acceptance before the *starting date* and afterwards as the *Project Manager* instructs.
(3) The *Project Manager* accepts the policies and certificates if they comply with the contract.
(4) If the *Contractor* does not submit a required policy or certificate, the *Employer* may insure a risk that the contract requires the *Employer* to insure.
(5) The *Contractor* pays the cost of this insurance to the *Employer.*
(6) The Parties comply with the terms and conditions of the insurance policies.

1.6.23 Disputes

Either the *Contractor* or the *Employer* may notify a dispute to the other Party. The procedure followed depends on whether the contract is subject to the Housing Grants, Construction and Regeneration Act 1996 and whether Option Y(UK)2 in ECC2 or Option W1 or W2 in ECC3 has been chosen as part of the contract strategy by the *Employer*.

Any dispute should first attempt to be sorted between the *Contractor* and the *Project Manager* before following the formal route of adjudication.

1.6.23.1 How the Project Manager *notifies a dispute*

(1) The *Project Manager* may not notify a dispute. The Parties to the contract – that is, the *Employer* or the *Contractor* – notify all disputes.

1.6.23.2 How the Employer *notifies a dispute*

(1) The *Employer* notifies a dispute to the *Contractor* at any time during the period of the contract.
(2) The procedure for adjudication depends on whether the contract is a construction contract under the Housing Grants, Construction and Regeneration Act 1996 and also whether Option Y(UK)2 has been chosen for the contract. In ECC3 the dispute clauses have been rewritten and the dispute resolution procedure depends on whether dispute resolution Option W1 where the HGCR Act 1996 does not apply or dispute resolution Option W2 where the HGCR Act 1996 does apply has been chosen.

1.6.23.3 How the Contractor *notifies a dispute*

(1) The *Contractor* notifies a dispute to the *Project Manager* at any time during the period of the contract.
(2) The procedure for adjudication depends on whether the contract is a construction contract under the Housing Grants, Construction and Regeneration Act 1996 and also whether Option Y(UK)2 has been chosen for the contract. In ECC3 the dispute clauses have been rewritten and the dispute resolution procedure depends on whether dispute resolution Option W1 where the HGCR Act 1996 does not apply or dispute resolution Option W2 where the HGCR Act 1996 does apply has been chosen.

1.6.24 Termination Either the *Employer* or the *Contractor* may terminate the contract. The ECC lists the reasons for which either Party may terminate, as well as the procedures and the payments on termination. The *Project Manager* may not terminate the contract; only the *Employer* or the *Contractor* may.

1.6.24.1 How the Employer *terminates the contract* The *Employer* may terminate for any reason, including those in the termination table included in clause 94.2, ECC3 clause 90.2. The procedures and the payments on termination differ depending on what reason is stated for termination.

(1) The *Employer* notifies the *Project Manager* that he wishes to terminate and gives the *Project Manager* his reasons for wanting to terminate.
(2) If the reason is R11, R12, R13, R14 or R15, the *Project Manager* should check the notification of the default to the *Contractor*, requiring the default to be rectified or stopped and that four weeks have passed since the notification. If the reason is R19 or R21 (R18 or R20 in ECC3), the *Project Manager* should check that 13 weeks have passed since an instruction to stop or not to start any substantial work without a further instruction allowing the work to restart or start.
(3) The *Project Manager* issues the termination certificate promptly if the reason given for termination complies with this contract.
(4) The *Contractor* does no further work necessary to complete the *works*.
(5) The *Employer* commences the procedures for termination immediately. The procedure differs with the reason for termination.
(6) Within 13 weeks of termination, the *Project Manager* certifies a final payment to or from the *Contractor*, which is the *Project Manager*'s assessment of the amount due on termination less the total of previous payments. The payment assessment differs with the reason for termination.

1.6.24.2 How the Contractor *terminates the contract* The *Contractor* may terminate for a reason included in the termination table included in clause 94.2, ECC3 clause 90.2 only. The procedures and the payments on termination differ depending on what reason is stated for termination.

(1) The *Contractor* notifies the *Project Manager* that he wishes to terminate and gives the *Project Manager* his reasons for wanting to terminate.
(2) If the reason is R16, the *Project Manager* should check the date of the certificate and that 13 weeks have passed since the certificate without payment. If the reason is R20 or R21 (R19 or R20 in ECC3), the *Project Manager* should check that 13 weeks have passed since an instruction to stop or not to start any substantial work without a further instruction allowing the work to restart or start.
(3) The *Project Manager* issues the termination certificate promptly if the reason given for termination complies with this contract.
(4) The *Contractor* does no further work necessary to complete the *works*.
(5) The *Contractor* and the *Employer* commence the procedures for termination immediately. The procedure differs with the reason for termination.
(6) Within 13 weeks of termination, the *Project Manager* certifies a final payment to or from the *Contractor*, which is the *Project Manager*'s assessment of the amount due on termination less the total of previous payments. The payment assessment differs with the reason for termination.

Pro-forma 23: TERMINATION CERTIFICATE

Project number:

Description:

Contract number:

To: (The *Contractor*)		To: (The *Project Manager*)	
Address:		Address:	
Telephone:		Telephone:	
Fax:		Fax:	
Attention:		Attention:	

The reasons for termination are:	
The procedure on termination is:	
Payments on termination are:	

	Day	Month	Year
The *Contractor*'s responsibility for the *works* ceases on:			

Certified by *Project Manager*:

Signature Name Date

Agreed by Legal Manager:

Signature Name Date

Distribution:				

Index

Terms in *italics* are identified in Contract Data and defined terms have capital initial letters.
Page numbers in *italics* refer to diagrams or illustrations.